Graphing Calculator Manual

For Tan's

Finite Mathematics
for the Managerial, Life, and Social Sciences
Seventh Edition

Yvette Hester and John Ryan

THOMSON

BROOKS/COLE

Australia • Canada • Mexico • Singapore • Spain • United Kingdom • United States

For more information about our products, contact us at:
Thomson Learning Academic Resource Center
1-800-423-0563

For permission to use material from this text, contact us by:
Phone: 1-800-730-2214
Fax: 1-800-731-2215
Web: www.thomsonrights.com

Asia
Thomson Learning
5 Shenton Way #01-01
UIC Building
Singapore 068808

Australia
Nelson Thomson Learning
102 Dodds Street
South Street
South Melbourne, Victoria 3205
Australia

Canada
Nelson Thomson Learning
1120 Birchmount Road
Toronto, Ontario M1K 5G4
Canada

Europe/Middle East/South Africa
Thomson Learning
High Holborn House
50-51 Bedford Row
London WC1R 4LR
United Kingdom

Latin America
Thomson Learning
Seneca, 53
Colonia Polanco
11560 Mexico D.F.
Mexico

Spain
Paraninfo Thomson Learning
Calle/Magallanes, 25
28015 Madrid, Spain

Contents

Introduction

The traditional finite mathematics course is ridden with cumbersome calculations which often obscure the underlying mathematical processes. But the use of graphing calculators can transform the traditional finite mathematics course into an understandable and enjoyable mathematical experience. The TI-83 series of graphing calculators arguably represent the most simple and powerful calculating devices for finite mathematics.

This manual is a self-contained introduction to the use of the TI-83/83 Plus graphing calculators for a finite mathematics course. It is not a tutorial in finite mathematics but an organized collection of examples worked entirely with the TI-83/83 Plus calculators. Our style is simple. Basic calculator commands are demonstrated with concrete examples rather than general syntax, and each example includes all of the relevant calculator screens.

There are no exercises in this manual since there is an ample supply of exercises included in Tan's Finite Mathematics textbook. Our hope is that students and instructors will use this manual as a source of worked practice problems which will allow them to gain confidence with the TI83/83 Plus as they attempt the associated exercises in Tan's Finite Mathematics textbook.

Comments regarding the differences in using the TI-83/83 Plus are given when necessary. A number of programs which give the TI-82 some of the improved features of the TI-83/83 Plus are included in chapter 10.

We would like to extend our sincere gratitude to Brooks/Cole for giving us the opportunity to produce this manuscript, and to Jeff Morgan for his endless patience helping us bring the camera-ready manuscript to its final form.

Yvette Hester, John Ryan

Chapter 1

Lines and Linear Functions

1.1 A Brief Introduction to the TI-83/83+

The home screen is where all basic calculations are performed. Start by making sure that the calculator is set to **Floating Point** mode. Press [**MODE**], next to the yellow 2nd key on the 83+, highlight **Float** using the arrow keys and press [**ENTER**].

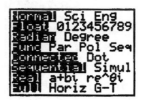

Press [**2nd**] [**MODE**] [**QUIT**]) to return to the 83+ home screen.

To calculate

$$2.1\left(\frac{3.1}{63.75} - 34.2\right) - 19.73$$

enter the quantities above (parenthesis keys are above the 8 and 9 keys) and press [**ENTER**].

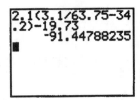

Note that multiplication is "implied". We get the same result if we include the multiplication sign in the calculation.

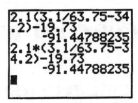

The placement of parentheses is very important. If we exclude the parentheses, the response from the TI-83/83+ is much different. Press [**CLEAR**] (underneath the arrow keys) to clear the home screen.

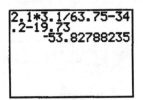

Round brackets must be used for parentheses. If either square or set brackets are used, the calculator gives an error message. These brackets are reserved for other purposes.

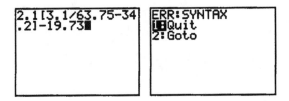

(Press [**ENTER**] to return to the home screen.)

Two important commands that can be accessed from the home screen are [**ENTRY**] ([**2nd**] ([**ENTER**]) and [**ANS**] ([**2nd**] [(−)]). The [**ENTRY**] command pastes the input from a previous calculation in the home screen. For example, if you have performed the calculations above, the following screens can be obtained by pressing [**ENTRY**] twice.

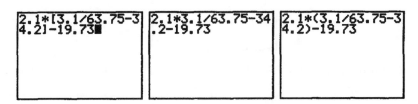

The [**ENTRY**] command can be used to correct mistakes by using the arrow keys to go to the mistake and retype, or to save time in entering calculations that are similar to ones that have already been performed.

The [**ANS**] command recalls the answer from the previous calculation. For example, if we compute $2.132(10.29 - 6.38)$ and then decide to divide this quantity by 23.52, we can press [**ANS**] [÷] **23.52** followed by [**ENTER**].

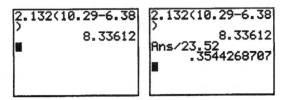

Four other useful commands are $[(-)]$, $[x^{-1}]$, $[x^2]$ and $[\wedge]$.

- $[(-)]$ negates a number (This is different from the operation of subtraction, which is a different key on the calculator.)
- $[x^{-1}]$ gives the reciprocal of a number
- $[x^2]$ squares a number
- $[\wedge]$ raises a number to a power

These commands can be combined to compute $-\left(\dfrac{1}{5} + 7.34^2\right)^3$ by pressing $[(-)]$

$[(]$ **5** $[x^{-1}]$ $[+]$ **7.34** $[x^2]$ $[)]$ $[\wedge]$ **3** [**ENTER**].

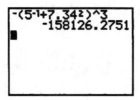

If $[-]$ (the "minus key") is pressed instead of $[(-)]$, on the 83+, we get an error message.

(Press [**ENTER**] to return to the home screen). An error message also appears if we try to use $[(-)]$ in place of $[-]$ to compute $5 - 3$.

This manual will illustrate calculator solutions to certain examples in Tan's Applied Finite Mathematics textbook, 7th edition.

The first is **Example 1 from Section 1.1:**

Find the distance between the points $(-4,3)$ and $(2,6)$.

Solution: Using the distance formula: $d = \sqrt{\left(2-(-4)\right)^2 + (6-3)^2}$.

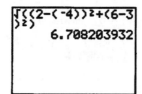

Variables

It is often necessary to keep track of a number and use it repeatedly in several calculations. In this case, it is convenient to assign a variable name to the number. We can assign a number to any variable name by using the [**STO⇒**] and [**ALPHA**] keys.

Note: Variable names can only be one character in length.

1.2 Graphing Lines and Setting the View Window

There are 2 steps associated with graphing lines on the TI-83/83+.

- Inputting the function in the **Y=** screen

- Defining the appropriate **GRAPH** window dimensions

Tan's Example 8 from Section 1.2:

Graph $y = 3x - 4$.

Solution: Press [**MODE**] and make sure **Func** is selected. Then press [**Y=**]. If any functions are currently present in this window, place the cursor to the right of each "= sign" and press [**CLEAR**] to remove them. Now enter the line in **Y1**.

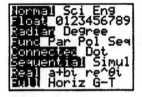 

To graph this line in the standard window (x between −10 and 10, y between −10 and 10), set the window by pressing [**WINDOW**] and filling in the values shown below, or [**ZOOM**] **6** will place you in this standard window. Press [**GRAPH**] to sketch the curve.

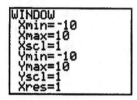 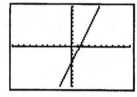

(If your axes do not appear, select **AxesOn** in the **FORMAT** screen.)

To access this line from the home screen, press [**VARS**], select **Y-VARS** and Function (use the right arrow key and press [**ENTER**]). Then select **Yl** by pressing [**ENTER**].

This will paste **Y1** (along with a left parenthesis) to the home screen. To compute *y* for *x* = 1.28, input **1.28** [)] [**ENTER**] to complete the calculation.

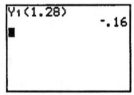

Note: It is not possible to access the function **Y1** by simply typing this quantity using the [**ALPHA**] button. If this is done, the TI-83/83+ assumes that we are requesting the product of **Y** "times" **1**.

Tan's Example 11 from Section 1.2:

Graph the line $y = 5000x + 50000$.

Solution: Enter the line in the [**Y=**] screen. Press [**GRAPH**] (the [**WINDOW**] screen is shown below for reference).

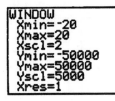 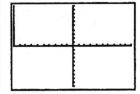

It appears as though only a piece of the graph lies in this window. This is a common problem. Some thought is required to obtain the proper view window. Note that the *y*-intercept of this line is 50000. Change the dimensions of the Graph window as shown below and press [**GRAPH**].

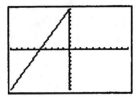

Finding Zeros

Example: Find the *x*-intercept of each of the lines above from the **GRAPH** window.

Solution: The *x*-intercept of the line $y = 5000x + 50000$ can be found using the **zero** command from within the [**CALC**] menu. Supply left and right bounds by using the arrow keys, and supply a guess (each followed by [**ENTER**]).

 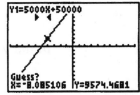 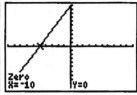

Sometimes it is easier to find the *x*-intercept of a line by simply solving the equation

$$5000x + 50000 = 0$$

for *x*. However, the zero command can be applied to more complicated functions where it is impossible (or impractical) to solve the corresponding equation by hand.

1.3 Evaluating Functions in the Graph Window

Tan's Example 3 from Section 1.3:

Evaluate $p = -\dfrac{1}{4}x + 20$ at $x = 40$ from the **GRAPH** window.

Solution: Graph the line as shown below.

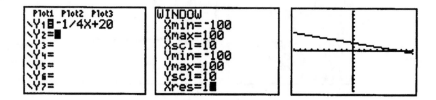

Now press [**CALC**] ([**2nd**] [**TRACE**]) and [**ENTER**] to select value. Enter **40** at the prompt for **X=,** and press [**ENTER**].

The TI-83/83+ places a marker at the corresponding point on the graph.

If you ask for a value that does not lie in the view window, the calculator will return an error message.

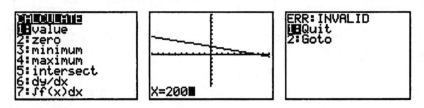

Now consider graphing more than one line in the same **GRAPH** window.

Example: Graph both $p = -\dfrac{1}{4}x + 20$ and $p = \dfrac{5}{4}x + 30$ and evaluate both at $x = 30$.

Solution: Enter the linear functions as **Y1** and **Y2**, with the [**WINDOW**] set as in the previous example and press [**GRAPH**].

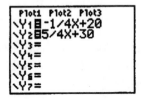   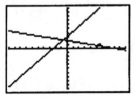

Both lines can be evaluated at $x = 30$ from the **GRAPH** window by pressing [**CALC**] [**ENTER**] (to select value). Enter the value **30** at the **X=** prompt and press [**ENTER**]. Use the up/down arrow keys to select the line to be evaluated at $x = 30$. The function value is listed at the bottom of the screen next to **Y=**.

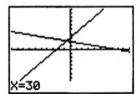   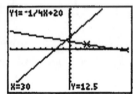   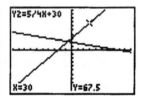

If more than 2 lines are being evaluated, continue to press the up/down arrow keys to select the desired line.

Trace

The **TRACE** command can be used to move the cursor along the graph of a function and give the coordinates of the corresponding points.

Tan's Example 4 from Section 1.3:

Graph the line $p = \dfrac{5}{4}x + 30$.

Solution: Place *p* in the [**Y=**] screen, set the **WINDOW** values and press [**GRAPH**].

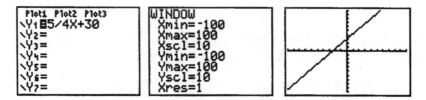

To use the **TRACE** command, make sure **CoordOn** has been selected in the [**FORMAT**] screen. The TI-83/83+ can access this screen directly by pressing [**2nd**] [**ZOOM**]. Now press [**GRAPH**] and then [**TRACE**].

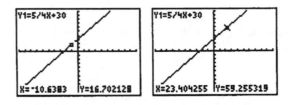

The right/left arrow keys can be used to move the cursor along the graph of the line. The coordinates of the points display at the bottom of the screen as the cursor moves along the graph.

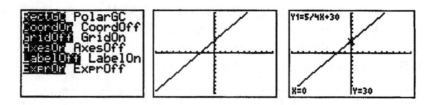

If more than one line is graphed, the up/down arrow keys can be used (after pressing [**TRACE**]) to select the desired graph.

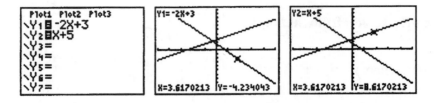

10

1.4 Intersection

Tan's Example 2 from Section 1.4:

Find the break-even point for $R(x) = 10x$ and $C(x) = 4x + 12{,}000$.

Solution: We can proceed in one of two ways. One approach is to use [**TRACE**] and [**ZOOM**] to approximate the point of intersection. A second approach is to use the **intersect** command from within the [**CALC**] menu.

Trace and Zoom

Let's begin by demonstrating [**TRACE**] and [**ZOOM**]. (Although you will eventually bypass this process when determining the intersection point of two lines, this process can be useful when a point of intersection of more complicated curves is required).

Start by using [**TRACE**] to obtain a reasonable approximation to the point of intersection. Then press [**ZOOM**] and select Zoom In. Press [**ENTER**] twice.

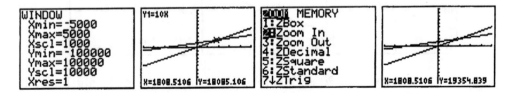

Now use [**TRACE**] again and repeat this process 2 more times. You should obtain the first two graphs below. The last graph is the result of several more applications of this process.

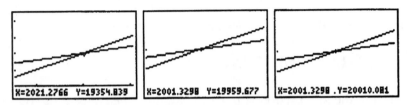

Consequently, the point of intersection is approximately (2001.3, 20010) .

Intersect

To obtain an accurate point of intersection, use the **intersect** command. Press [**CALC**], select item 5 and press [**ENTER**]. (The point of intersection must appear in your window.)

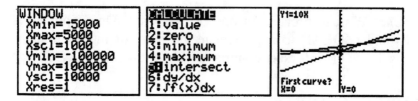

The TI-83/83+ prompts for the first curve. This is because there might be several curves plotted, with several intersection points. Use the up/down arrow keys to select the first curve and press [**ENTER**] (if only two curves are present, just press [**ENTER**]). Repeat this process for the second curve. The calculator prompts for a guess. Use the right/left arrow keys to trace to the approximate point of intersection and press [**ENTER**].

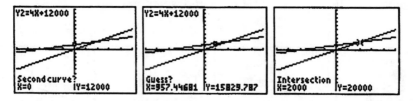

The point of intersection is (2000, 20000) . This can be confirmed algebraically by setting the following equations equal to each other.

$$\left(\begin{array}{l} R = 4x + 12,000 \\ C = 10x \end{array} \right)$$

A summary of bulleted items for intersection follows.

- Press [**CALC**] ([**2nd**] [**TRACE**]).

- Select 5 (the intersection you are looking for *must* appear in your graph window).

- Place cursor on first graph. Press [**ENTER**].

- Place cursor on second graph. Press [**ENTER**].

- Trace to get close to intersection. Press [**ENTER**].

- The coordinates of the point of intersection will appear at the bottom of the screen.

1.5 Least Squares

Lists

Lists can be created and viewed by using the **STAT Editor**. Press [**STAT**] [**ENTER**] to edit lists **Ll, L2, ...**

To input numbers into **L1**, use the arrow keys to move to the first entry below the heading **Ll** and press (−) 1.6 [**ENTER**]. Then enter the additional values −4.9, −14.2, −24.13, −32.2.

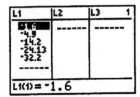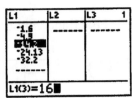

To input numbers into the list **L2**, use the arrow keys to move to the first entry below the heading **L2** and press 5 [**ENTER**]. Then enter the additional values 1, 2.474, −25, 124, 34.

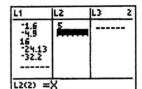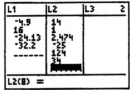

To clear **L2**, from the **STAT Editor** highlight the **L2** heading, press [**CLEAR**], and press the down arrow key.

L1	**L2**	L3	2
-1.6	5	------	
-4.9	14		
16	1		
-24.13	2.474		
-32.2	-25		
------	124		
	34		

L2 = {5, 14, 1, 2.47...

L1	**L2**	L3	2
-1.6	5	------	
-4.9	14		
16	1		
-24.13	2.474		
-32.2	-25		
------	124		
	34		

L2 =

L1	L2	L3	2
-1.6	▬▬▬	------	
-4.9			
16			
-24.13			
-32.2			

L2(1)=

Least Squares

Least squares is the process of finding the line that gives the least squares approximation for a given set of data points. This line is typically referred to as the *least squares line or regression line.*

Tan's Example 1 from Section 1.5:

Find the least squares line for the data

$$P_1(1,1), P_2(2,3), P_3(3,4), P_4(4,3), P_5(5,6)$$

Solution: Press [**STAT**] [**ENTER**] and input the values for x and y in **L1** and **L2**.

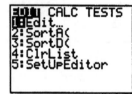

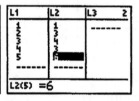

Press [**Y=**] and clear any functions. (Or simply turn them off by un-highlighting the equal sign. Use the arrow keys to do this and press enter.) Now press [**STAT PLOT**] ([**2nd**] [**Y=**]) [**ENTER**] . Make the selections shown in the third screen below.

 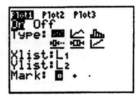

To obtain the scatter plot, [**ZOOM**] **9** (the calculator puts the data in an appropriate statistics window) or set window dimensions as below and press [**GRAPH**].

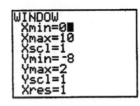 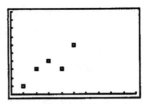

To obtain the least squares line for this set of data points, press [**STAT**] and use the arrow keys to highlight **CALC**. Select **LinReg(ax+b)** and press [**ENTER**] twice.

 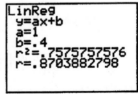

(To see the correlation coefficient **r** on the TI-83/83+, select **Diagnostics On** under catalog.)

To obtain a graph of this line among the scatter plot of data points, press [**Y=**], then [**VARS**] and select Statistics. Select **EQ** and press [**ENTER**] to paste the least squares line into **Y1**.

The final [**Y=**] screen and the plot obtained by pressing [**GRAPH**] are shown below.

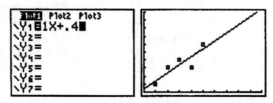

A summary of bulleted items for the procedure follows.

- Enter data into the lists using [**STAT**] [**EDIT**].

- Clear [**Y=**] screen. Make the appropriate choices in the [**STAT PLOT**] window using up/down arrows.

- Set dimensions for the **GRAPH** window or [**ZOOM**] **9**.

- Select **LinReg(ax+b)** under **CALC** in the [**STAT**] menu. Press [**ENTER**] twice.

- Paste the least squares line into **Y1** by pressing [**Y=**] and then [**VARS**].

- Select Statistics and press [**ENTER**]. Select **EQ** and press [**ENTER**].

- [**GRAPH**] to obtain the graph of the least squares line among the data points.

The Appendix contains a program named **LSQ** which can be used to automate most of this process.

Chapter 2

Systems of Equations and Matrices

Matrix operations on the TI-83/83+ can be used to solve linear systems of equations. This chapter introduces basic matrix operations on the TI-83/83+ and demonstrates these operations in solving linear systems of equations.

2.1 Introduction to Systems

Tan's Example 1 from Section 2.1:

Consider the system
$$2x - y = 1$$
$$3x + 2y = 12$$

Solving each equation for y, we get
$$y = 2x - 1$$
$$y = -\frac{3}{2}x + 6$$

Graphing these two lines yields the following windows.

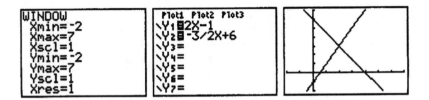

Using **the intersect** command we get the following sequence of windows:

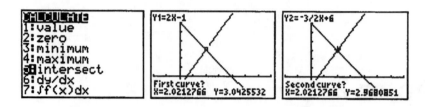

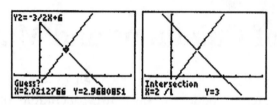

Therefore, the unique solution occurs at (2,3).

Tan's Example 2 from Section 2.1:

Consider the system

$$2x - y = 1$$
$$6x - 3y = 3$$

Solving each equation for y, we get

$$y = 2x - 1$$
$$y = 2x - 1$$

Graphing these two lines yields the following windows.

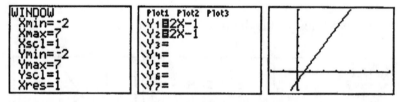

Since the lines are coincident, the system has infinitely many solutions.

Tan's Example 3 from Section 2.1:

Consider the system

$$2x - y = 1$$
$$6x - 3y = 12$$

Solving each equation for y, we get

$$y = 2x - 1$$
$$y = 2x - 4$$

Graphing these two lines yields the following windows.

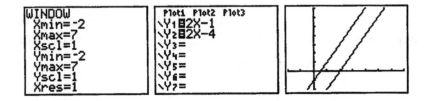

The two lines are parallel and never meet. Therefore, the system has no solutions.

2.2 Solving Systems of Linear Equations I

Matrices are easy to create and store on the TI-83/83+.

Tan's Example 2 from Section 2.2:

To enter the matrix

$$\begin{pmatrix} 2 & 4 & 6 & 22 \\ 3 & 8 & 5 & 27 \\ -1 & 1 & 2 & 2 \end{pmatrix}$$

press [2nd] [x⁻¹] (If you are using the TI-83 press the [MATRX] button), highlight **EDIT** and select one of the *preset* matrix names. We select [A] and press [ENTER].

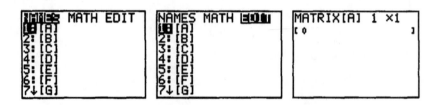

Now we need to input the dimensions of the matrix. The matrix above is 3 x 4. This information is given to the calculator by pressing **[3] [ENTER] [4] [ENTER]**. A matrix appears and the numbers can be entered, each followed by **[ENTER]**. When the input process is finished, **[QUIT]** to the home screen.

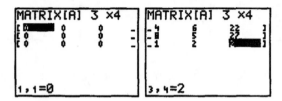

To recall the matrix **[A]** to the home screen and use it in calculations, press **[MATRX]**, highlight the name of the desired matrix and press **[ENTER]**. This writes the matrix *name* to the home screen. Press **[ENTER]** again to view the matrix.

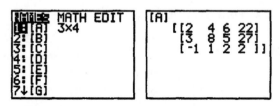

(Note that it is not possible to recall the matrix by simply typing square brackets with the letter **A** enclosed.)

Using Gauss-Jordan and Pivoting to Solve Linear Systems

Three basic calculator commands will be used repeatedly in this section. They are shown below along with their locations in the calculator.

- **rowSwap** (obtain by pressing [**MATRX**], highlighting **MATH** and scrolling, if necessary)

- ***row** (obtained in the same manner as rowSwap)

- ***row+** (obtained in the same manner as rowSwap)

We demonstrate the use of these commands to implement the Gauss-Jordan and pivoting in the example below.

Tan's Example 5 from Section 2.2:

Use the Gauss-Jordan method to solve the system
$$\begin{pmatrix} 3x - 2y + 8z = 9 \\ -2x + 2y + z = 3 \\ x + 2y - 3z = 8 \end{pmatrix}$$

Solution: Enter the augmented matrix for this system as [**A**] and display it on the home screen.

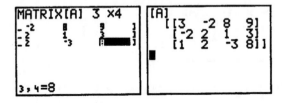

To begin the Gauss-Jordan process, the (row 1 column 1) entry must be 1.

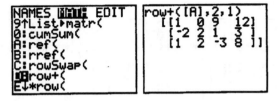

We can eliminate the (row 2, column 1) and (row 3, column 1) entries by using the ***row+** command.

```
[-2 2 1  3]
[1  2 -3 8]]
*row+(2,Ans,1,2)

[[1 0 9  12]
 [0 2 19 27]
 [1 2 -3  8]]
```

```
[0 2 19 27]
[1 2 -3  8]]
*row+(-1,Ans,1,3)
)
[[1 0 9  12]
 [0 2 19 27]
 [0 2 -12 -4]]
■
```

Note that the command *row+(2,Ans,1,2) in the first screen above tells the TI-83/83+ to add **2** times **row 1** to **row 2** and place the result in **row 2**. Similarly, the command *row+(-1,Ans,1,3) in the second screen above tells the TI-83/83+ to add **-1** times **row 1** to **row 3** and place the result in **row 3**.

Now swap **row 2** with **row 3**.

```
[0 2 19  27]
[0 2 -12 -4]]
rowSwap(Ans,2,3)

[[1 0 9  12]
 [0 2 -12 -4]
 [0 2 19  27]]
■
```

Now multiply **row 2** by 1/2 and eliminate the (row 3, column 2) entry.

```
[[1 0 9  12]
 [0 2 -12 -4]
 [0 2 19  27]]
*row(1/2,Ans,2)
[[1 0 9  12]
 [0 1 -6 -2]
 [0 2 19 27]]
```

```
[0 1 -6 -2]
[0 2 19 27]]
*row+(-2,Ans,2,3)
)
[[1 0 9  12]
 [0 1 -6 -2]
 [0 0 31 31]]
```

(Note that the [**ENTRY**] command can be very helpful here. Recall that it will repeat the last entry and allow editing.) Finally, multiply **row 3** by 1/31 and then eliminate the (row 1, column 3) and (row 2, column 3) entries.

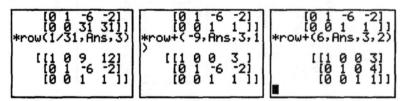

```
[0 1 -6 -2]
[0 0 31 31]]
*row(1/31,Ans,3)

[[1 0 9  12]
 [0 1 -6 -2]
 [0 0 1   1]]
```

```
[0 1 -6 -2]
[0 0 1  1]]
*row+(-9,Ans,3,1)
)
[[1 0 0  3]
 [0 1 -6 -2]
 [0 0 1  1]]
```

```
[0 1 -6 -2]
[0 0 1   1]]
*row+(6,Ans,3,2)

[[1 0 0 3]
 [0 1 0 4]
 [0 0 1 1]]
■
```

Consequently, the solution to the original system is given by $x = 3$, $y = 4$, $z = 1$ (3, 4, 1).

The TI-83/83+ has a built in command for automating the process above. The rref command is accessed by pressing [**MATRX**], highlighting **MATH** and scrolling to **rref**. The result of applying this command to the augmented matrix for the first example in this section is shown below.

The appendix contains a program RowOps to automate the process if each step needs to be recorded.

2.3 Solving Systems of Linear Equations II

Recall that systems may also have infinitely many or no solutions.

Tan's Example 1 from Section 2.3:

Solve the system of equations given by
$$\begin{cases} x + 2y - 3z = -2 \\ 3x - y - 2z = 1 \\ 2x + 3y - 5z = -3 \end{cases}$$

Solution: The individual steps associated with performing the Gauss-Jordan method are shown below. Enter the augmented matrix **A** and proceed, using the three allowable commands.

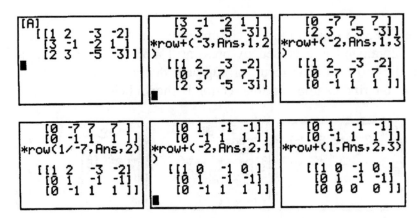

The final screen tells us that the original system is equivalent to

$$\begin{pmatrix} x - z = 0 \\ y - z = -1 \end{pmatrix}$$

Solving for x and y in terms of z gives

$$x = z \text{ and } y = z - 1$$

with z being a free parameter.

So, the solution is given by $(z, z - 1, z)$, where z is any real number. Therefore, the original system has infinitely many solutions.

Using the **rref** command (accessed by pressing [**MATRX**], highlighting **MATH** and scrolling to **rref**) the result for the previous example is shown below.

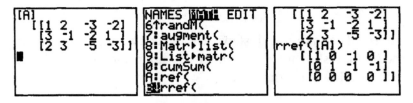

Tan's Example 2 from Section 2.3:

Solve the system of linear equations given by

$$\begin{pmatrix} x+y+z=1 \\ 3x-y-z=4 \\ x+5y+5z=-1 \end{pmatrix}$$

Solution: We apply the Gauss-Jordan process to this system as shown below.

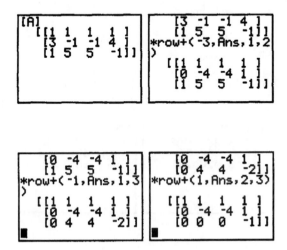

The last screen shows that the original system is equivalent to the system

$$\begin{pmatrix} x+y+z=1 \\ -4y-4z=1 \\ 0=-1 \end{pmatrix}$$

Since the last equation is not valid, this system has no solution.

2.4 More on Matrices

Two matrices of the same dimensions can added or subtracted.

Tan's Example 3 from Section 2.4:

Suppose

$$A = \begin{pmatrix} 320 & 280 & 460 & 280 \\ 480 & 360 & 580 & 0 \\ 540 & 420 & 200 & 880 \end{pmatrix}$$

and

$$B = \begin{pmatrix} 210 & 180 & 330 & 180 \\ 400 & 300 & 450 & 40 \\ 420 & 280 & 180 & 740 \end{pmatrix}$$

Enter A and B into the appropriate matrices. To find $A + B$, follow the windows below.

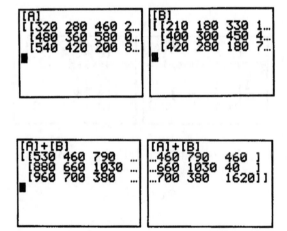

Tan's Example 5 from Section 2.4:

Using B from the previous example

$$B = \begin{pmatrix} 210 & 180 & 330 & 180 \\ 400 & 300 & 450 & 40 \\ 420 & 280 & 180 & 740 \end{pmatrix}$$

suppose we want to multiply matrix B by 1.1. Type **[1] [.] [1] [2nd] [x⁻¹]** **[ENTER] [⇓] [ENTER]**

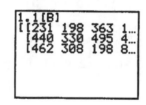

or simply type **[1] [.] [1] [2nd] [ANS] [ENTER]** as is shown in the following two screens.

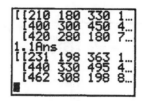

We can store this new matrix into matrix **[C]** by using the **[STOP]** command. Press **[2nd] [ANS] [STOP] [MATRX]**, highlight **[C]** and press **[ENTER]** twice.

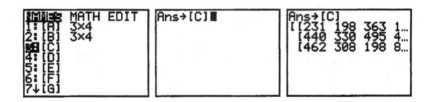

2.5 Matrix Multiplication

It is very easy to multiply matrices with the TI-83/83+.

Tan's Example 3 from Section 2.5:

Enter the matrices

$$A = \begin{pmatrix} 3 & 1 & 4 \\ -1 & 2 & 3 \end{pmatrix} \text{ and } B = \begin{pmatrix} 1 & 3 & -3 \\ 4 & -1 & 2 \\ 2 & 4 & 1 \end{pmatrix}$$

The product of [A] times [B] can be computed from the home screen.

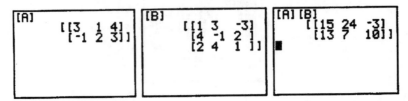

Note that the product of [B] times [A] is not valid, and the TI-83/83+ responds with an error message.

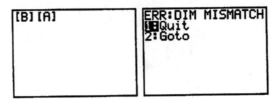

A summary of bulleted items for entering and multiplying matrices follows.

- Press [**MATRX**] [**EDIT**] and select a matrix name. Press [**ENTER**].

- Input number of rows, [**ENTER**], number of columns, [**ENTER**].

- Enter each number in the matrix, followed by [**ENTER**].

- [**QUIT**] to the home screen.

To multiply two matrices (check your dimensions):

- From the home screen, press [**MATRX**], select desired matrix, press [**ENTER**].

- Press [**MATRX**], select desired matrix, press [**ENTER**] [**ENTER**].

2.6 Using Inverses to Solve Linear Systems

Most linear systems have the same number of unknowns as equations. These systems yield coefficient matrices that are square.

Tan's Example 1 from Section 2.6:

Find the inverse of the matrix

$$A = \begin{pmatrix} 2 & 1 & 1 \\ 3 & 2 & 1 \\ 2 & 1 & 2 \end{pmatrix}$$

Solution: Input the matrix [A] as shown below.

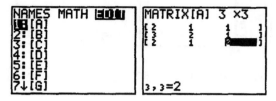

Now **[QUIT]** to the home screen and press **[MATRX]** **[ENTER]** **[x⁻¹]** **[ENTER]**.

Consequently,

$$A^{-1} = \begin{pmatrix} 3 & -1 & -1 \\ -4 & 2 & 1 \\ -1 & 0 & 1 \end{pmatrix}$$

NOTE: *Not all matrices have inverses.* The process given above will fail if the coefficient matrix, A, does not have an inverse.

Tan's Example 2 from Section 2.6:

Find the inverse of the matrix

$$A = \begin{pmatrix} 1 & 2 & 3 \\ 2 & 1 & 2 \\ 3 & 3 & 5 \end{pmatrix}$$

Solution: Input the matrix [A] as shown below

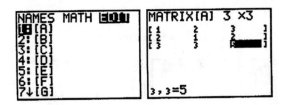

Now [QUIT] to the home screen and press [MATRX] [ENTER] [x⁻¹] [ENTER].

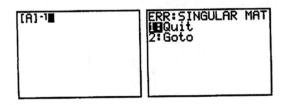

Consequently, A^{-1} does not exist.

Tan's Example 4a from Section 2.6:

Solve the linear system

$$2x + y + z = 1$$
$$3x + 2y + z = 2$$
$$2x + y + 2z = -1$$

by using an inverse matrix.

Solution: The system can be rewritten in the form

$$\begin{pmatrix} 2 & 1 & 1 \\ 3 & 2 & 1 \\ 2 & 1 & 2 \end{pmatrix} \begin{pmatrix} x \\ y \\ z \end{pmatrix} = \begin{pmatrix} 1 \\ 2 \\ -1 \end{pmatrix}$$

$$A \qquad X = \quad B$$

and the solution is given by

$$\begin{pmatrix} x \\ y \\ z \end{pmatrix} = \begin{pmatrix} 2 & 1 & 1 \\ 3 & 2 & 1 \\ 2 & 1 & 2 \end{pmatrix}^{-1} \begin{pmatrix} 1 \\ 2 \\ -1 \end{pmatrix}$$

$$X = \quad A^{-1} \qquad B$$

provided $\begin{pmatrix} 2 & 1 & 1 \\ 3 & 2 & 1 \\ 2 & 1 & 2 \end{pmatrix}^{-1}$ exists.

To perform this calculation on the TI-83/83+, enter the matrices [A] and [B] as shown below.

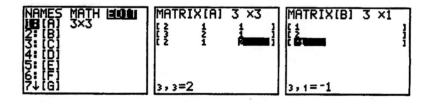

[QUIT] to the home screen and press
[MATRX] [ENTER] [x⁻¹] [MATRX] [B] [ENTER] [ENTER]

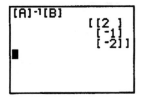

Consequently, the solution to the system above is given by

$$x = 2, \; y = -1, \; z = -2 \quad (2, -1, -2).$$

2.7 Leontief Input-Output Model

Tan's Example 2 from Section 2.7:

We need to solve the matrix equation $X = (I - A)^{-1} D$, where

$$A = \begin{pmatrix} .2 & .2 & .1 \\ .2 & .4 & .1 \\ .1 & .2 & .3 \end{pmatrix} \text{ and } D = \begin{pmatrix} 100 \\ 80 \\ 50 \end{pmatrix}$$

Solution: Input the matrices [**I**], [**A**] and [**D**] as shown below.

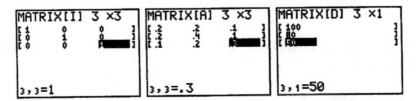

Now compute $X = (I - A)^{-1} D$.

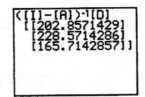

Chapter 3

A Geometric Approach to Linear Programming

This chapter demonstrates the use of reverse shading and the **intersect** command to geometrically solve linear programming problems in two variables.

3.1 Reverse Shading Linear Inequalities

To geometrically identify a solution region, we use reverse shading. Reverse shading is the process of shading the side of a line that we are NOT INTERESTED IN. If we repeat this process for all of the lines which define a solution set, then the **unshaded region** will be the solution region.

Every linear inequality is related to an associated line. We consider two types of lines below.

- Lines which are not vertical
- Vertical lines

Reverse Shading Non Vertical Lines

Tan's Example 1 from Section 3.1:

Example 1: *Reverse shade* the inequality

$$2x + 3y \geq 6$$

Solution: The first step is to rewrite the inequality in terms of y, giving

$$y \geq -\frac{2}{3}x + 2$$

To reverse shade this inequality on the **TI-83/83+**, enter the line $y = -\frac{2}{3}x + 2$ as **Y1** and set appropriate **WINDOW** dimensions.

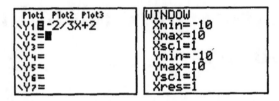

Since the inequality for y above involves "$\geq$", the reverse shading process will shade *below* the line (always confirm the region with a test point). Use the left arrow key to highlight the line symbol next to the equation name in the [**Y=**] window. Use the [**ENTER**] key to toggle between the choices. Select the upper triangle to shade above a line. Select the lower triangle to shade below the line. Press [**GRAPH**].

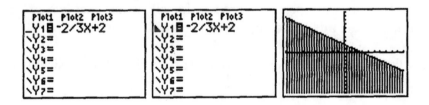

Reverse Shading Vertical Lines

Tan's Example 2 from Section 3.1:

We have to use a *trick* to get the TI-83/83+ to shade one side of a vertical line.

Example 3: *Reverse shade* the inequality
$$x \leq -1$$

Solution: In this case, *reverse shading* corresponds to shading the right side of the line $x = -1$. Here's the trick. First, input a line with EXTREMELY LARGE NEGATIVE slope which passes through the point $(-1, 0)$. This line will approximate the line $x = -1$. The choice $y = -10^9(x - (-1))$ seems to work best.

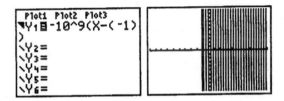

Then we can shade the right hand side of this line on the **TI-83/83+** as shown below.

NOTE: Using negative slope above is critical. Otherwise, the shading will occur on the opposite side of the vertical line.

Tan's Example 3 from Section 3.1:

Reverse shade the inequality
$$x - 2y > 0$$

Solution: Again, the inequality must first be solved for y, giving
$$y < \frac{1}{2}x.$$

Since the inequality sign above is "<", reverse shading results in shading *above* the graph of the line $y < \frac{1}{2}x$ (don't forget to use a test point). The reverse shading process on the **TI-83/83+** is shown below.

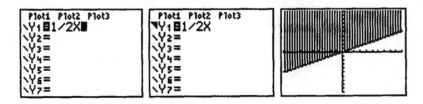

Once again, the unshaded portion of the graph (not including the line) corresponds to the solution set of

$$x - 2y > 0$$

Tan's Example 4 from Section 3.1:
Determine the solution set for the system

$$4x + 3y \geq 12$$
$$x - y \leq 0$$

Solution: Solve each inequality for y, giving

$$y \geq -\frac{4}{3}x + 4$$
$$y \geq x$$

Since the inequality sign above is "$\geq$", reverse shading results in shading *below* the graph of the lines $y \geq -\frac{4}{3}x + 4$ and $y \geq x$. The process on the **TI-83/83+** is shown below.

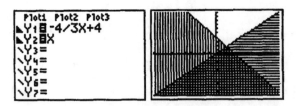

The unshaded portion of the graph (including the bordering lines) corresponds to the solution set of

$$y \geq -\frac{4}{3}x + 4$$

$$y \geq x$$

Tan's Example 5 from Section 3.1:

Determine the solution set for the system

$$x \geq 0$$

$$y \geq 0$$

$$x + y - 6 \leq 0$$

$$2x + y - 8 \leq 0$$

Solution: Rewrite these inequalities so that the last three are given in terms of y.

$$x \geq 0$$

$$y \geq 0$$

$$y \leq -x + 6$$

$$y \leq -2x + 8$$

Then make the following four entries in the [**Y=**] screen.

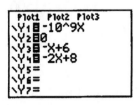

The inequalities are "$\leq$" for lines 3 and 4, so *reverse shading* corresponds to shading above these lines, whereas the inequalities are "$\geq$" for lines 1 and 2, so *reverse shading* corresponds to shading below these lines (check with test points). On the **TI-83/83+**, simply toggle the symbols to the left of the equation names in the [**Y=**] window as shown below and press [**GRAPH**].

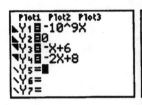 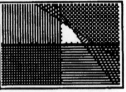

The unshaded portion of the graph (including the bordering lines) corresponds to the solution set of

$$x \geq 0$$
$$y \geq 0$$
$$x + y - 6 \leq 0$$
$$2x + y - 8 \leq 0$$

3.2 Reverse Shading For Larger Systems

Tan's Example 2 from Section 3.2:

Determine the solution set for the system

$$40x + 10y \geq 2400$$
$$10x + 15y \geq 2100$$
$$5x + 15y \geq 1500$$
$$x \geq 0$$
$$y \geq 0$$

Solution: Rewrite these inequalities so that the first three are given in terms of *y*.

$$y \geq -4x + 240$$

$$y \geq -\frac{2}{3}x + 140$$

$$y \geq -\frac{1}{3}x + 100$$

$$x \geq 0$$

$$y \geq 0$$

Then make the following five entries in the [**Y=**] screen and resize the window.

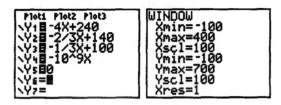

The inequalities are "$\geq$" for the lines, so *reverse shading* corresponds to shading below these lines (check with test points). On the **TI-83/83+**, simply toggle the symbols to the left of the equation names in the [**Y=**] window as shown below and press [**GRAPH**].

The unshaded portion of the graph (including the bordering lines) corresponds to the solution set of

$$40x + 10y \geq 2400$$

$$10x + 15y \geq 2100$$

$$5x + 15y \geq 1500$$

$$x \geq 0$$

$$y \geq 0$$

Tan's Example 3 from Section 3.2:

Determine the solution set for the system

$$x+y \geq 40$$
$$x+y \leq 100$$
$$x \leq 80$$
$$y \leq 70$$
$$x \geq 0$$
$$y \geq 0$$

Solution: Rewrite these inequalities so that the first, second, fourth and fifth are given in terms of *y*.

$$y \geq -x+40$$
$$y \leq -x+100$$
$$x \leq 80$$
$$y \leq 70$$
$$x \geq 0$$
$$y \geq 0$$

Then make the following six entries in the [Y=] screen and resize the window.

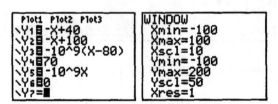

The inequalities are "$\leq$" for lines 2, 3 and 4, so *reverse shading* corresponds to shading above these lines, whereas the inequalities are "$\geq$" for lines 1, 5 and 6, so *reverse shading* corresponds to shading below these lines (check with test points). On the **TI-83/83+**, simply toggle the symbols to the left of the equation names in the [Y=] window as shown below and press [**GRAPH**].

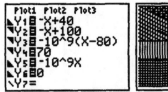
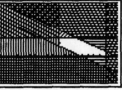

The unshaded portion of the graph (including the bordering lines) corresponds to the solution set of

$$x + y \geq 40$$
$$x + y \leq 100$$
$$x \leq 80$$
$$y \leq 70$$
$$x \geq 0$$
$$y \geq 0$$

3.3 The Graphical Solution of a Linear Programming Problem

Tan's Example 1 from Section 3.3:

Maximize $\quad P = x + 1.2y$
subject to $\quad 2x + y \leq 180$
$\qquad\qquad x + 3y \leq 300$
$\qquad\qquad x \geq 0$
$\qquad\qquad y \geq 0$

Solution: The feasible region associated with these constraints is shown below.

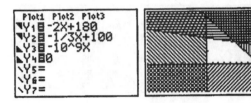

Use the [TRACE] command and the arrow keys to determine which lines intersect to form the corner points above. The screens below indicate that the top corner point is the intersection of **Y1** with **Y2**.

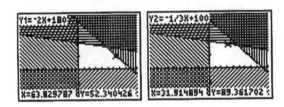

Use the **intersect** command to find the top corner point. Press [CALC] (from the screen above) and select intersect. The calculator will prompt for the first and second curves. Use the up/down arrow keys (followed by [ENTER]) to input **Y1** and **Y2** respectively.

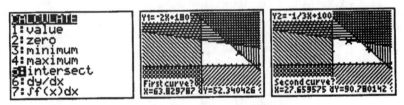

The calculator will now prompt for a guess. Press [ENTER].

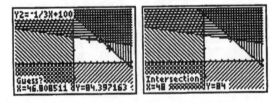

So the upper corner point is (3,8). Now [QUIT] to the home screen. Enter **X + 1.2Y** and [ENTER] to evaluate the objective function at this corner point. The

intersect command stored the coordinates of the corner point in **X** and **Y**, respectively.

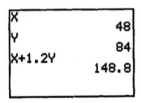

Press **X [ENTER]** and **Y [ENTER]** to illustrate.

The other three corner points can be found in a similar fashion. They are given by (0, 0) , (90, 0) and (0,100). The last screen from each of the processes is shown below.

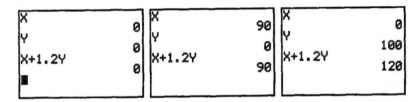

From these calculations, the maximum value of the objective function occurs at (48, 84) with a maximum value of 148.80.

Chapter 4

The Simplex Method

The first section in this chapter shows how to use the built-in matrix operations to perform the simplex method on a standard maximum problem. The second section shows how to use the program **PIVOT** which can be found in the appendix on programming, to solve a standard minimization problem.

4.1 Row Operations and the Simplex Method

The example below gives the basic ideas behind performing the simplex method on the TI-83/83+.

Tan's Example 3 from Section 4.1:

Maximize $P = 2x + 2y + z$
subject to

$$2x + y + 2z \leq 14$$
$$2x + 4y + z \leq 26$$
$$x + 2y + 3z \leq 28$$
$$x \geq 0, y \geq 0, z \geq 0$$

Solution: The first step is to form the simplex table

$$\begin{pmatrix} 2 & 1 & 2 & 1 & 0 & 0 & 0 & 14 \\ 2 & 4 & 1 & 0 & 1 & 0 & 0 & 26 \\ 1 & 2 & 3 & 0 & 0 & 1 & 0 & 28 \\ -2 & -2 & -1 & 0 & 0 & 0 & 1 & 0 \end{pmatrix}$$

Store this matrix. Write it to the home screen and scroll the entries to make sure we have entered it correctly.

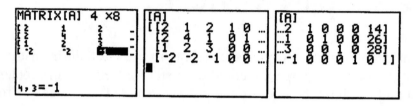

Since the most negative entry in the last row (−2) occurs twice, we may choose either the first or the second column as the pivot column. We will select the first column. By checking the ratios of entries (last column divided by first column), we determine that the pivot entry is in (row 1, column 1). The screens below show the results of multiplying row 1 by 1/2, and using row operations to eliminate the entries in (row 2, column 1), (row 3, column 1) and (row 4, column 1).

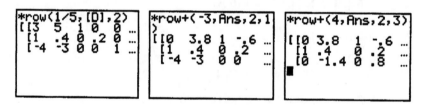

If we scroll right in the last screen, we see that the new simplex table is given by

$$\begin{pmatrix} 1 & .5 & 1 & .5 & 0 & 0 & 0 & 7 \\ 0 & 3 & -1 & -1 & 1 & 0 & 0 & 12 \\ 0 & 1.5 & 2 & -.5 & 0 & 1 & 0 & 21 \\ 0 & -1 & 1 & 1 & 0 & 0 & 1 & 14 \end{pmatrix}$$

The new pivot column is 2 and the pivot entry lies in (row 2, column 2). The results of pivoting on this entry axe shown below.

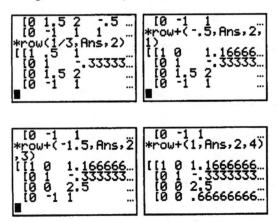

Convert the resulting matrix into fraction form

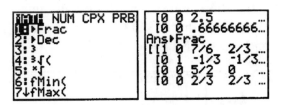

Scrolling right shows that the new simplex table is given by

$$\begin{pmatrix} 1 & 0 & 7/6 & 2/3 & -1/6 & 0 & 0 & 5 \\ 0 & 1 & -1/3 & -1/3 & 1/3 & 0 & 0 & 4 \\ 0 & 0 & 5/2 & 0 & -1/2 & 1 & 0 & 15 \\ 0 & 0 & 2/3 & 2/3 & 1/3 & 0 & 1 & 18 \end{pmatrix}$$

All the entries in the last row are nonnegative, so we have reached the optimal solution. We conclude that

$$x = 5, \ y = 4, \ z = 0,$$
$$P = 18, \ u_1 = 0, u_2 = 0, u_3 = 15$$

4.2 The Simplex Method: Standard Minimization Problem

Now let's use the **PIVOT** program from the Appendix to solve a standard minimization problem.

Example 1 in Section 4.2:

Solution: Start by storing the simplex table

$$\begin{pmatrix} 5 & 4 & 1 & 0 & 0 & 32 \\ 1 & 2 & 0 & 1 & 0 & 10 \\ -2 & -3 & 0 & 0 & 1 & 0 \end{pmatrix}$$

in matrix **[D]**. Since the **PIVOT** program specifically refers to **[D]**, this matrix name must be used. To run the **PIVOT** program, press **[PRGM]**, scroll to highlight **PIVOT**, and press **[ENTER]**.

Press **[ENTER]** and the following menu screen appears.

Define the pivot entry first, then press **[ENTER]** give the row and column numbers of the pivot entry followed each time by **[ENTER]**.

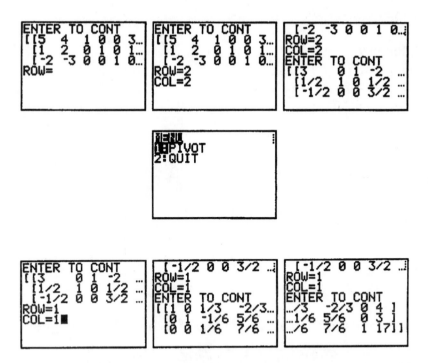

If we scroll right, we find that we have reached the final simplex table shown below.

$$\begin{pmatrix} 1 & 0 & 1/3 & -2/3 & 0 & 4 \\ 0 & 1 & -1/6 & 5/6 & 0 & 3 \\ 0 & 0 & 1/6 & 7/6 & 1 & 17 \end{pmatrix}$$

As in the last section, we find that P has the minimum value of 17 when $x = 4$, $y = 3$, $u_1 = 0, u_2 = 0$.

4.3 The Simplex Method: Nonstandard Problems

Tan's Example 4 from Section 4.3:

To solve

Minimize	$C = 2x - 3y$
subject to	$x + y \le 5$
	$x + 3y \ge 9$
	$-2x + y \le 2$
	$x \ge 0,\ y \ge 0$

consider the following tableau:

$$\begin{pmatrix} 1 & 1 & 1 & 0 & 0 & 0 & 5 \\ -1 & -3 & 0 & 1 & 0 & 0 & -9 \\ -2 & 1 & 0 & 0 & 1 & 0 & 2 \\ 2 & -3 & 0 & 0 & 0 & 1 & 0 \end{pmatrix}$$

Begin by storing this matrix into [**D**] and use the **PIVOT** program.

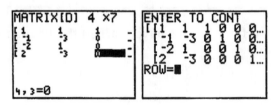

Using our rules for nonstandard maximization problems we determine that the pivot element is (row 3 column 2).

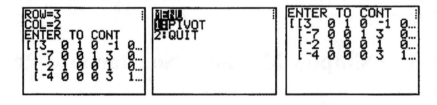

The next pivot element is (row 2 column 1).

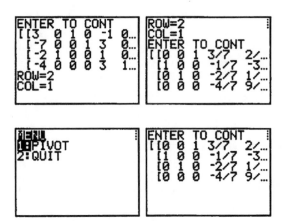

The final pivot element is (row 1 column 4).

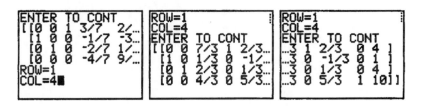

The final tableau is

$$\begin{pmatrix} 0 & 0 & 7/3 & 1 & 2/3 & 0 & 4 \\ 1 & 0 & 1/3 & 0 & -1/3 & 0 & 1 \\ 0 & 1 & 2/3 & 0 & 1/3 & 0 & 4 \\ 0 & 0 & 4/3 & 0 & 5/3 & 1 & 10 \end{pmatrix}$$

and the optimal solution is $x = 1$, $y = 4$, and $C = -P = -10$.

Chapter 5

Mathematics of Finance

The TI-83/83+ **TVM Solver** can be used to tackle problems dealing with compound interest, annuities, amortization and sinking funds. We have included a TI-82 program, **TVM**, in the Appendix.

5.1 The Time Value of Money (TVM) Solver

This section will discuss the meaning of the variables found on the **TVM Solver** screen and the sign conventions when dealing with cash flows.

To access the **TVM Solver** press [Apps] [**ENTER**] ([2nd] [Finance] and [**ENTER**] for the TI-83)

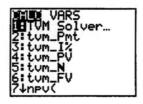 

- **N** is the number of payment periods $(n \cdot t)$

- **I %** is the annual interest rate

- **PV** is the present value

- **PMT** is the payment amount

- **FV** is the future value

- **P/Y** is the number of payments per year

- **C/Y** is the number of conversion periods per year

- **PMT: END BEGIN** is when the payments occur in a payment period

Sign conventions are that cash inflows are positive and cash outflows are negative. For example, if we borrow $5000 from a bank and agree to pay the bank $100 a month for 5 years, then *PV* = 5000 and *PMT* = −100.

Simple Interest

Tan's Example 1 from Section 5.1:

$$P = 1000, r = .08, t = 3$$

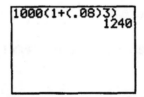

Note: The TVM Solver can only be used for simple interest problems when $t = 1$.

Tan's Example 3(d) from Section 5.1:

Find the accumulated amount after 3 years if $1000 is invested at 8% compounded monthly.

Solution: In this case, we have

- $N = 3 * 12 = 36$

- $I\% = 8$
- $PV = -1000$ (it is an outflow)
- $P/Y = 12$
- $C/Y = 12$

To solve this problem, go to the **TVM Solver**. Press **[APPS] [ENTER] [ENTER]** (**[2nd] [FINANCE] [ENTER]** for the TI-83) and input the values into the calculator as shown below.

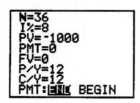

To find the future value, place the cursor to the right of **FV=** and press **[ALPHA] [ENTER]** to solve for **FV**.

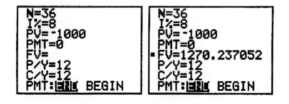

So, the accumulated value is $1270.24 after 3 years.

Effective Interest Rate

To compute the effective interest rate, given the nominal interest rate and the number of compounding periods, we use the ►**Eff** function (Choice C in the finance menu). This command takes two arguments, and has the form

$$\blacktriangleright \textbf{Eff(nr,M)}$$

where

- **nr** is the nominal rate
- **M** is the number of times compounding occurs in a period

Tan's Example 4(d) from Section 5.1:

Find the effective rate of an investment with a nominal rate of 8% compounded monthly.

Solution: Press [**2nd**] [**FINANCE**], scroll down to ▶**Eff**, press [**ENTER**] and input the values shown below.

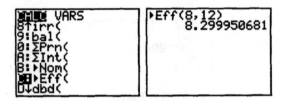

So the effective rate of the investment is approximately 8.3%.

NOTE: The nominal rate can be computed from a known effective rate by using the ▶**Nom** command.

Present Value

Tan's Example 5 from Section 5.1:

How much money should be deposited in a bank paying interest at the rate of 6% per year compounded monthly so that at the end of three years the accumulated amount will be $20,000?

Solution: In this case, we have

- $N = 3 * 12 = 36$

- $I\% = 6$

- $FV = 20000$

- $P/Y = 12$

- $C/Y = 12$

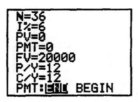

To find the present value, place the cursor to the right of **PV=** and press [**ALPHA**] [**ENTER**] to solve for **PV**.

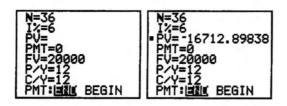

So, the present value is approximately $16,713.

5.2 Annuities and Sinking Funds

The financial functions on the TI-83/83+ allow us to easily perform computations associated with annuities and sinking funds.

Tan's Example 1 from Section 5.2:

Find the amount of an ordinary annuity of 12 monthly payments of $100 that earn interest at 12% per year compounded monthly.

Solution: In this case, we have

- $N = 12$

- $I\% = 12$

- $PMT = -100$ (it is an outflow)

- $P/Y = 12$

- $C/Y = 12$

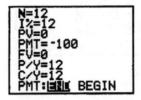

To find the future value, place the cursor to the right of **FV=** and press [**ALPHA**] [**ENTER**] to solve for **FV**.

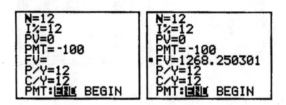

So, the future value of the annuity is approximately $1268.25.

Tan's Example 2 from Section 5.2:

Find the present value of an ordinary annuity of 24 payments of $100 each made monthly and earning interest at 9% per year compounded monthly.

Solution: In this case, we have

- $N = 24$

- $I\% = 9$

- $PMT = -100$ (it is an outflow)

- $P/Y = 12$

- $C/Y = 12$

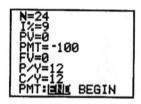

To find the present value, place the cursor to the right of **PV=** and press [**ALPHA**] [**ENTER**] to solve for **PV**.

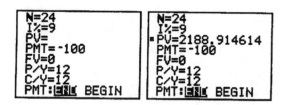

So, the present value of the annuity is approximately $2188.92.

Tan's Example 4 from Section 5.2:

After making a down payment of $2000 for an automobile, Murphy paid $200 per month for 36 months with interest charged at 12% per year compounded monthly on the unpaid balance. What was the original cost of the car?

Solution:

Original cost of the car = Present Value of the loan + Down Payment. We will first compute the present value of the loan. In this case, we have

- $N = 36$

- $I\% = 12$

- $PMT = -200$ (it is an outflow)

- $P/Y = 12$

- $C/Y = 12$

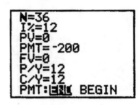

To find the present value, place the cursor to the right of **PV=** and press [**ALPHA**] [**ENTER**] to solve for **PV**.

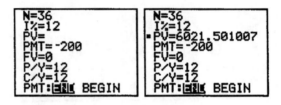

The present value of the loan is $6021.50. So, the original cost of the car is $6021.50 + $2000 = $8021.50.

5.3 Amortizations and Sinking Funds
Amortizations

Tan's Example 1 from Section 5.3:

A sum of $50, 000 is to be repaid over a 5-year period through equal installments made at the end of each year. If an interest rate of 8% per year is charged on the unpaid balance and interest calculations are made at the end of each year, determine the size of each installment so that the loan is amortized at the end of 5 years.

Solution: In this case, we have

- $N = 5$

- $I\% = 8$

- $PV = 50000$ (it is an inflow)

- $P/Y = 1$

- $C/Y = 1$

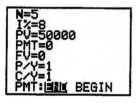

To find the payments, place the cursor to the right of **PMT=** and press [**ALPHA**] [**ENTER**] to solve for **PMT**.

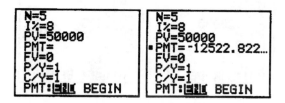

So, the annual payments are $12,522.82.

Tan's Example 2 from Section 5.3:

The Blakelys borrowed $120,000$ from a bank to help finance the purchase of a house. The bank charges interest at the rate of 9% per year on the unpaid balance, with interest computations being made at the end of each month. The Blakelys have agreed to repay the loan in equal monthly installments over the next 30 years. How much should each payment be if the loan is amortized at the end of the term?

Solution: In this case, we have

- $N = 30 * 12 = 360$

- $I\% = 9$

- $PV = 120000$ (it is an inflow)

- $P/Y = 12$

- $C/Y = 12$

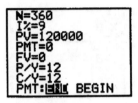

To find the payments, place the cursor to the right of **PMT=** and press [**ALPHA**] [**ENTER**] to solve for **PMT**.

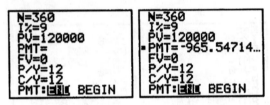

So, the monthly payments are $965.55.

Sinking Funds

Tan's Example 5 from Section 5.3:

The proprietor of Carson Hardware has decided to set up a sinking fund for the purpose of purchasing a computer in 2 years' time. It is expected that the computer will cost $30, 000. If the fund earns 10% interest per year compounded quarterly, determine the size of each (equal) quarterly installment the proprietor should pay into the fund.

Solution: In this case, we have

- $N = 4*2 = 8$

- $I\% = 10$

- $FV = -30000$ (it is an inflow)

- $P/Y = 4$

- $C/Y = 4$

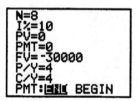

To find the payments, place the cursor to the right of **PMT=** and press [**ALPHA**] [**ENTER**] to solve for **PMT**.

So, the monthly payments are $3434.02.

5.4 Arithmetic and Geometric Progressions

Before we get started we need to become familiar with two new commands:

- **seq**

- **sum**

To create a sequence of numbers in the form of a list we use the **seq** function. This command takes five arguments, and has the form

seq(expression, variable, begin, end, increment).

If we wanted to create the terms in the arithmetic progression 2, 2+5, 2+2*5, ... 2+19*5 we would input
$$seq(2 + (N - 1) * 5, N, 1, 20, 1).$$

To access the **seq** function press [**2nd**] [**STATS**] highlight **OPS** and select **seq** as shown in the following windows.

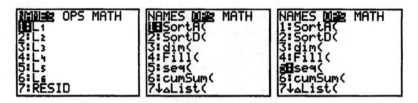

Finishing the expression we get the following windows.

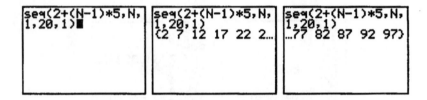

To sum the elements in a list we use the **sum** function.

$$sum(\textbf{list})$$

Suppose we wanted to sum the terms in the sequence
$$7, 15, 23, \ldots 87.$$

To access the **sum** function press [**2nd**] [**STATS**] highlight **MATH** and select sum as shown below.

Now we create the list using the **seq** command and complete the computation.

```
NAMES ▣▨ MATH    sum(seq(N,N,7,87    sum(seq(N,N,7,87
1:SortA(         ,8))                ,8))
2:SortD(                                            517
3:dim(
4:Fill(
▣seq(
6:cumSum(
7↓⌐List(
```

Arithmetic Progressions

Tan's Example 3 from Section 5.4:

Find the sum of the first 20 terms in the arithmetic progression 2, 2+5, 2+2∗5, ... 2+19∗5.

Solution: To access the **sum** function press **[2nd] [STATS]** highlight **MATH** and select **sum** as shown below.

Now we create the list using the **seq** command and complete the computation as shown below.

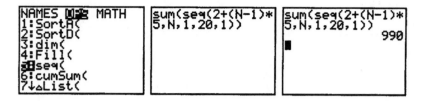

So, the sum of the first 20 terms in the arithmetic progression 2, 2+5, 2+2∗5, ... 2+19∗5 is 990.

Geometric Progressions

Tan's Example 7 from Section 5.4:

Find the sum of the first 6 terms in the geometric progression $3, 3*2$, $3*2^2$, ...

Solution: To access the **sum** function press **[2nd] [STATS]** highlight **MATH** and select **sum** as shown below.

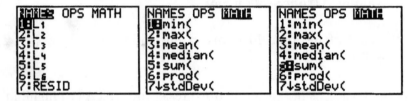

Now we create the list using the **seq** command and complete the computation as shown below.

So, the sum of the first six terms in the geometric progression $3, 3*2$, $3*2^2$, ... is 189.

Chapter 6

Counting

Factorials

The TI-83/83+ can be used to compute factorials. The command is found by pressing [**MATH**], highlighting **PRB** and scrolling to choice 4.

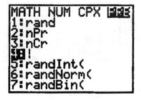

Tan's Example 2 from Section 6.4:

Find the number of ways a baseball team consisting of nine people can arrange themselves in a line for a group picture.

Solution: Applying the multiplication principle, we conclude that there are 9! arrangements.

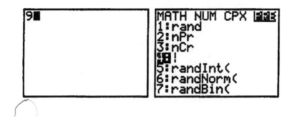

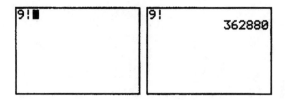

So, there are 362,800 ways the baseball team can be arranged for the picture.

Permutations

The TI-83/83+ can be used to compute permutations. The command is found by pressing [**MATH**], highlighting **PRB** and scrolling to choice 2. **nPr** is used to compute the number of permutations of *n* objects taken *r* at a time.

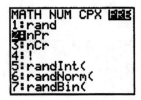

Tan's Example 5 from Section 6.4:

Find the number of ways a chairman, a vice-chairman, a secretary, and a treasurer can be chosen from a committee of eight members.

Solution: The problem is equivalent to finding $P(8,4)$

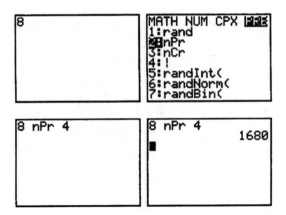

So, there are 1,680 ways of choosing the four officials from the committee of eight members.

Combinations

The TI-83/83+ can be used to compute combinations. The command is found by pressing [**MATH**], highlighting **PRB** and scrolling to choice 3. **nCr** is used to compute the number of combinations of *n* objects taken *r* at a time.

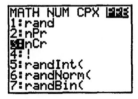

Tan's Example 5 from Section 6.4:

A state senate investigation subcommittee of four members is to be selected from a Senate committee of ten members. Determine the number of ways this can be done.

Solution: The problem is equivalent to finding $C(10, 4)$.

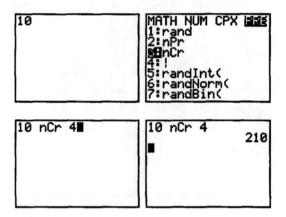

So, there are 210 ways of choosing such a committee.

Chapter 7

Probability

Combinations from Chapter 6 can be combined with the concept of probability.

Tan's Example 2 from Section 7.4:

Two cards are selected at random from a well-shuffled pack of 52 playing cards. What is the probability that:

a. They are both aces?

b. Neither of them is an ace?

Solution: For part a,

$$P(E) = \frac{n(E)}{n(S)} = \frac{C(4,2)}{C(52,2)}$$

So, the probability that both cards selected are aces is approximately .00452.
If we want to convert this number to a fraction we can press [**MATH**] [**ENTER**] [**ENTER**] as shown in the following screens.

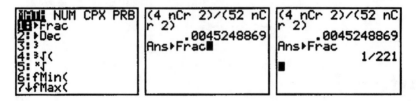

For part b,

$$P(E) = \frac{n(E)}{n(S)} = \frac{C(48,2)}{C(52,2)}$$

So, the probability that neither card selected is an ace is $\frac{188}{221} \approx .85$.

Tan's Example 1 from Section 7.6:

The picture tubes for the Pulsar 19-inch color television sets are manufactured in three locations and then shipped to the main plant of the Vista Vision Corporation for final assembly. Plants A, B, and C supply 50%, 30%, 20%, respectively, of the picture tubes used by Vista Vision. The quality-control department of the company has determined that 1% of the picture tubes produced by plant A are defective, whereas 2% of the picture tubes produced by plants B and C are defective. If a Pulsar 19-inch television set is selected at random and the picture tube is found to be defective, what is the probability that the picture tube was manufactured in plant C?

Solution: $P(C|D) = \dfrac{P(C \cap D)}{P(D)}$, so we need to compute the numerator and denominator of this fraction.

$$P(C \cap D) = P(C) \cdot P(C|D) = .2(.02)$$

We will store this value as the variable N as shown in the screen below.

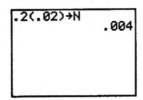

Now we need to compute the denominator.
$$P(D) = P(A) \cdot P(D|A) + P(B) \cdot P(D|B) + P(C) \cdot P(D|C)$$
$$= .5(.01) + .3(.02) + .2(.02)$$

We will store this value as the variable D as shown in the screens below.

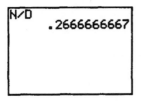

Taking the ratio N/D

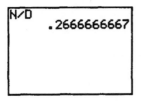

we find that the answer is approximately .27.

Chapter 8

Statistics

8.1 Probability Distributions and Histograms

We recall the lists material from section 1.5 to help us create histograms for probability distributions.

Tan's Example 5 from Section 8.1:

We will use the probability distribution for the sum of the faces of two dice.

x	2	3	4	5	6	7	8
$P(X = x)$	1/36	2/36	3/36	4/36	5/36	6/36	5/36

9	10	11	12
4/36	3/36	2/36	1/36

We begin by entering the data into **L1** and **L2** as shown below.

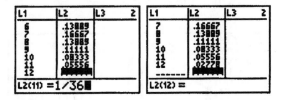

Press [**STAT PLOT**] ([**2nd**] [**Y=**]) [**ENTER**] and select the appropriate settings as shown below.

 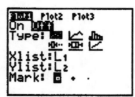 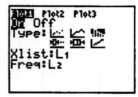

Now, press [**GRAPH**] to look at the histogram.

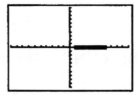

What a great histogram! Obviously, the window dimensions are not correct.

Window dimensions can either be set by entering appropriate values in the window menu or using **ZoomStat**. Press [**ZOOM**], highlight **ZoomStat** and press [**ENTER**].

 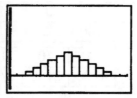

Note: ZoomStat doesn't always work. If it hadn't worked, we need to modify the window manually as shown in the screens below.

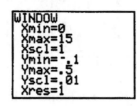 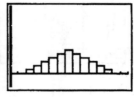

In Summary: To create a histogram

- Clear out **Y=** (or turn them off).

- Press **[STATPLOT] [ENTER]**.

- Arrow down and over to make bar graph selection. **[ENTER]**.

- Xlist: Ll **[ENTER]**.

- Freq/Prob: L2 **[ENTER]**.

- **[GRAPH]**.

8.2 Expected Value

The TI-83/83+ can be used to compute one variable statistics.

Tan's Example 2 from Section 8.2:

x	0	1	2	3	4	5	6	7	8
$P(X = x)$	.03	.15	.27	.20	.13	.10	.07	.03	.02

Solution: Store the data and probabilities in lists **Ll** and **L2**, respectively. Press **[STAT]**, highlight **CALC**, press **[ENTER]** and press **Ll [,] L2**.

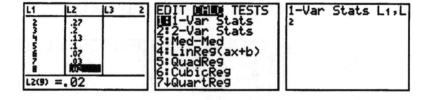

Now, press **[ENTER]** to get the one variable statistics and scroll to see the information.

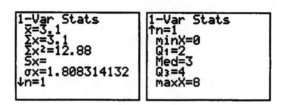

The mean (expected value) is given by

$$\overline{x} = 3.1$$

and the median is given by

$$Med = 3.$$

8.3 Variance and Standard Deviation

The one variable statistics list also gives us the standard deviation. We can then square that value to compute variance.

Tan's Example 1 from Section 8.3:

x	1	2	3	4	5	6	7
$P(X = x)$	.05	.075	.2	.375	.15	.1	.05

Solution: Store the data and probabilities in lists **Ll** and **L2**, respectively. Press [**STAT**], highlight **CALC**, press [**ENTER**] and press **Ll** [**,**] **L2**.

Now, press [**ENTER**] to get the one variable statistics and scroll to see the information.

The standard deviation is given by

$$\sigma_x = 1.3964$$

and we compute the variance by squaring the standard deviation (σ_x^2)

$$\sigma_x^2 = 1.95.$$

8.4 Binomial Distribution

The TI-83/83+ has a built-in probability distributions menu. The **binompdf** function can be used to compute the probabilities for the binomial distribution. This command takes three arguments, and has the form

binompdf(n, p, x)

where

- **n** is the number of trials

- **p** is the probability

- **x** is the number of successes.

Tan's Example 1 from Section 8.4:

A fair die is cast four times. Compute the probability of obtaining exactly one 6 in the four throws.

Solution: In this case, we have

- $n = 4$

- $p = 1/6$

- $x = 1$

To compute the probability press [2nd] [VARS] highlight **binompdf** and press [ENTER]

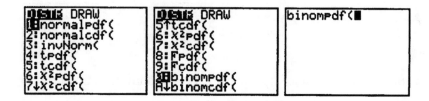

then input the values for n, p, and x as shown below.

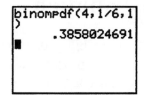

We can use the **binompdf** function to create an entire probability distribution by listing out all the values of x.

$$\textbf{binompdf}\left(n, p, \{x_1, x_2, ..., x_k\}\right)$$

Tan's Example 2(a) from Section 8.4:

A fair die is cast five times. If a 1 or a 6 lands uppermost in a trial, then the throw is considered a success. Find the probability distribution of obtaining exactly 0, 1, 2, 3, 4, and five successes, respectively, in this experiment.

Solution: To compute the probability distribution press [2nd] [VARS] highlight **binompdf** and press [ENTER]

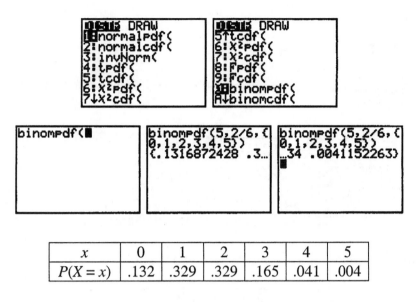

x	0	1	2	3	4	5
$P(X = x)$	.132	.329	.329	.165	.041	.004

We can use **binomcdf** function to find a cumulative probability (at most or no more than problems).

$$\textbf{binomcdf } (n, p, m)$$

- **n** is the number of trials

- **p** is the probability

- **m** is the largest number of successes

Tan's Example 5 from Section 8.4:

A division of the Solaron Corporation manufactures photovoltaic cells to use in the company's solar energy converters. It is estimated that 5% of the cells manufactured axe defective. If a random sample of 20 is selected from a large lot of cells manufactured by the company, what is the probability that it will contain at most 2 defective cells?

Solution: In this case, we have

- $n = 20$

- $p = .05$

- $m = 2$.

To compute the probability press [2nd] [VARS] highlight **binomcdf** and press [ENTER]

then input the values for n, p, and m as shown below.

We can use the **binomcdf** to compute the probability that at least m successes occur with

$$1 - \textbf{binomcdf}(n, p, m - 1).$$

Tan's Example 6(c) from Section 8.4:

The probability that a heart transplant performed at the Medical Center is successful is .7. Of six patients who have recently undergone such an operation what is the probability that at least three transplants are successful?

Solution: In this case, we have

- $n = 6$

- $p = .7$

- $m = 3$.

We must compute

$$1 - \textbf{binomcdf}(6, .7, 2).$$

To compute the probability press [**2nd**] [VARS] highlight **binomcdf** and press [**ENTER**]

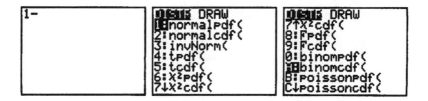

then input the values for n, p, and m as shown below.

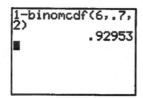

So, the answer is approximately .93.

8.5 The Normal Distribution

In this section we use the TI-83/83+ to compute z-scores and the area under the normal curve. We have included a TI-82 program, **ZTAB**, in the Appendix which can be used to perform all of the calculations in this section.

Computing Areas From Values of z

We will use **normalcdf** from the distributions menu to compute the area under the normal curve. The **normalcdf** command is used in the form

normalcdf(lowerbound, upperbound, mean, standard deviation)

and can be found by pressing [**2nd**] [**VARS**] [**2**].

Note: If the arguments mean and standard deviation are omitted when using **normalcdf**, the calculator will default to the values associated with a standard normal distribution ($\mu = 0$ and $\sigma = 1$).

Tan's Example 1 from Section 8.5:

Let Z be a standard normal variable. Find the following:

a. $P(Z < 1.24)$

b. $P(Z > 0.5)$

c. $P(0.24 < Z < 1.48)$

d. $P(-1.65 < Z < 2.02)$

Solution: For part a, we need to compute $P(Z < 1.24)$ which is the same as $P(-\infty < Z < 1.24)$. As a result, we need a negative number which is essentially $-\infty$ for the lower bound. The value -10^{99} is close enough. Press [**2nd**] [**VARS**], highlight **normalcdf**, press [**ENTER**] and input the values as shown below.

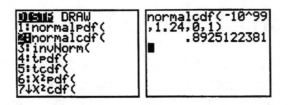

Consequently,
$$P(Z < 1.24) = 0.8925122381.$$

For part b, we need to compute $P(Z > 0.5)$ which is the same as $P(0.5 < Z < \infty)$. Using the same logic as above, we choose 10^{99} to play the role of ∞. Press [**2nd**] [**VARS**], highlight **normalcdf**, press [**ENTER**] and input the values as shown below.

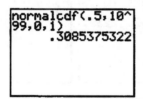

Therefore,

$$P(.5 < Z) = 0.3085375322.$$

For part c, press **[2nd]** **[VARS]**, select **normalcdf**, press **[ENTER]** and input the values shown below.

```
normalcdf(.24,1.
48,0,1)
        .3357285187
```

Consequently,

$$P(.24 < Z < 1.48) = 0.8610785742.$$

For part d, press **[2nd]** **[VARS]**, select **normalcdf**, press **[ENTER]** and input the values shown below.

```
normalcdf(-1.65,
2.02,0,1)
        .9288369247
```

Consequently,

$$P(.24 < Z < 1.48) = 0.9288369247.$$

Note: The standard normal distribution is used in the example above. Hence, it is not necessary to specify the mean and the variance. Parts a, b, c, and d are solved below using this shortcut.

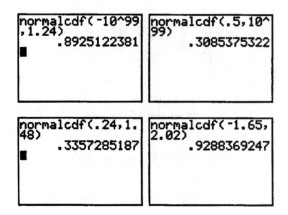

Using the Area to Compute z

The **invNorm** command can be used to compute the *z* value associated with an area under a normal curve to the left of the *x* value. The syntax for the **invNorm** command is given by

invNorm(area, mean, standard deviation).

Note: The calculator will default to the values associated with a standard normal distribution ($\mu = 0$ and $\sigma = 1$) whenever the arguments *mean* and *standard deviation* are omitted from the **invNorm** command.

Tan's Example 2(a & b) from Section 8.5:

Let *Z* be a standard normal variable. Find the value of *z* such that

a. $P(Z < z) = .9474$.

b. $P(Z > z) = .9115$.

Solution: For part a, **press [2nd] [VARS]** select **invNorm** and input the values as shown below.

The value shown above implies
$$z = 1.620150274.$$

For part b, $P(Z > z) = .9115$ is the same as finding $P(Z < -z) = .9115$. Press [2nd] [VARS] select **invNorm** and input the values as shown below.

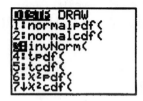
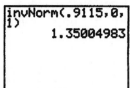

The value shown above implies
$$-z = 1.35004983.$$

So, $z = -1.35004983$.

Tan's Example 3 from Section 8.5:

Suppose X is a normal random variable with $\mu = 100$ and $\sigma = 20$. Find the following:

a. $P(X < 120)$

b. $P(X > 70)$

c. $P(75 < X < 110)$

Solution: For part a, we need to compute $P(X < 120)$ which is the same as $P(-\infty < X < 120)$. As a result, we need a negative number which is essentially $-\infty$ for the lower bound. The value -10^{99} is close enough. Press

[2nd] [VARS], highlight **normalcdf**, press [ENTER] and input the values as shown below.

Consequently,

$$P(X < 120) = 0.8413447404.$$

For part b, we need to compute $P(X > 70)$ which is the same as $P(70 < X < \infty)$. Using the same logic as above, we choose 10^{99} to play the role of ∞. Press [2nd] [VARS], highlight **normalcdf**, press [ENTER] and input the values as shown below.

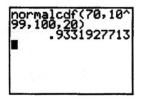

Therefore,

$$P(70 < X) = 0.9331927713.$$

For part c, press [2nd] [VARS], select **normalcdf**, press [ENTER] and input the values shown below.

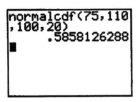

Consequently,

$$P(75 < X < 110) = 0.8610785742.$$

8.6 Applications of the Normal Distribution

Tan's Example 1 from Section 8.6:

The medical records of infants delivered at the Kaiser Memorial Hospital show that the infants' birth weights in pounds are normally distributed with a mean of 7.4 and a standard deviation of 1.2. Find the probability that an infant selected at random from among those delivered at the hospital weighted more than 9.2 pounds at birth.

Solution: We need to find $P(X > 9.2)$,

```
normalcdf(9.2,10
^99,7.4,1.2)
          .0668072287
■
```

So, $P(X > 9.2) = .0668072287$.

Tan's Example 5 from Section 8.6:

The probability that a heart transplant performed at the Medical Center is successful (that is, the patient survives 1 year or more after surgery) is .7. Of 100 patients who have undergone such an operation, what is the probability:

a. Fewer than 75 will survive 1 year or more after the operation?

b. Between 80 and 90, inclusive, will survive 1 year or more after the operation?

Solution: $\mu = 100(.7) = 70$ and $\sigma = \sqrt{100(.7)(.3)} = 4.58$. So, for part a,

$$P(X < 75) \approx P(Y < 74.5)$$

Y is normally distributed with mean 70 and standard deviation 4.58.

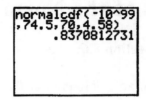

So, $P(X < 75) \approx .837$.

For part b,
$$P(80 \le X \le 90) \approx P(79.5 < Y < 90.5)$$

Y is normally distributed with mean 70 and standard deviation 4.58.

So, $P(80 \le X \le 90) \approx .019$.

Chapter 9

Markov Chains

In this chapter we will use the **TI-83/83+** to help solve two typical problems involving Markov chains.

Tan's Example 1 from Section 9.2:

A survey conducted by the National Commission on the Educational Status of Women reveals that 70% of the daugthers of women who have completed two or more years of college have also completed two or more years of college, whereas 20% of the daughters of women who have less than two years of college have completed two or more years of college. If this trend continues, determine, in the long run, the percentage of women in the population who will have completed at least two years of college given that currently only 20% of the women have completed at least two years of college.

Solution: The transition matrix for showing this information is

$$T = \begin{bmatrix} .7 & .3 \\ .2 & .8 \end{bmatrix}$$

and

$$X_0 = \begin{bmatrix} .2 & .8 \end{bmatrix}$$

After 1 generation

$$X_1 = X_0 T$$

Input the matrices X_0 and T as [**A**] and [**B**], respectively.

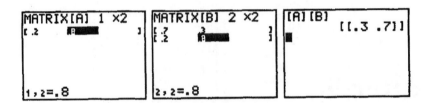

Therefore, after one generation 30% of the women have completed two or more years of college. After 10 generations

So, after ten generations 39.98% of the women have completed two or more years of college. Consequently, it appears that in the long run approximately 40% of the women will complete two or more years of college.

Tan's Example 3 from Section 9.2:

Find the steady-state distribution vector for the regular Markov chain whose transition matrix is

$$T = \begin{bmatrix} .7 & .3 \\ .2 & .8 \end{bmatrix}$$

Solution: The easiest way to solve this type of problem is to raise the transition matrix to a very high power (100) and read off the values from the columns of the new matrix. Enter the transition matrix into the calculator and raise it to a large power as in the following screens.

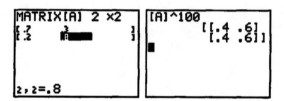

The steady-state distribution vector is of the form $\begin{bmatrix} x & y \end{bmatrix}$, where $x = .4$ and $y = .6$.

Appendix

Basic Programming Instructions and Programs for Customizing the TI-82/83/83+

This chapter includes an introduction to programming the TI-82/83/83+ calculator. In addition, it contains a few programs which allow the TI-82 to perform computations which are exclusive to the TI-83/83+.

The first section introduces programming on the TI-82/83/83+ through a simple program named **SLOPE**. The remaining sections contain simple programs which can be used on either the TI-82 or TI-83/83+. The programs in some of the latter sections are intended to give the TI-82 some of the added capabilities of the TI-83/83+. The programs in this chapter are listed below:

- **SLOPE** - a simple program which finds the slope of a line

- **COIN** - a simple program which simulates flipping coins

- **DICE** - a simple program which simulates rolling dice

- **LSQ** - automated least squares

- **RREF** - automated Gauss-Jordan for the TI-82

- **PIVOT** - a step by step simplex method

- **TVM** - a program for computing the time value of money for the TI-82

- **ZTAB** - a program for finding the area under the normal curve and computing z scores

An Introduction to Programming the TI-82/83/83+

Writing programs on the TI-82/83/83+ is a simple matter. A program is nothing more than a list of commands for the calculator to execute.

Let's write a program to illustrate the basic process. Suppose we want a program which can be used to find the slope of a line segment connecting two points. To start, press [**PRGM**], select [**NEW**] and press [**ENTER**].

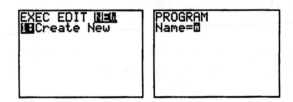

The calculator will prompt us for a name for our program. The calculator is already in **ALPHA-locked** mode, so we can start entering the name without worrying about pressing [**ALPHA**]. Let's name our program **SLOPE**. After we enter the name, press [**ENTER**], and the calculator sends us to the program editor with a blank program listing.

Before we proceed, we need to decide how our program will work. We want to have the calculator ask the user for the coordinates of two points on the line of interest, and then have the calculator compute and display the slope of the line. The program listing is shown below. The discussion following the listing explains how to enter each line of the program.

```
PROGRAM:SLOPE
:Disp "FIRST POINT"
:Input "X=",U
:Input "Y=",V
```

```
:Disp "SECOND POINT"
:Input "X=",X
:Input "Y=",Y
:If X=U
:Then
:Disp "VERTICAL LINE"
:Else
:Disp "SLOPE= ",(Y - V)/(X - U)
:End
```

Look closely at the program listing above. If you read it carefully (line by line), you should be able to understand the basic process (regardless of whether or not you have written a program before).

Now, let's enter this program. Do not attempt to type the commands **Disp**, **Input**, **If**, **Then**, **Else** and **End** using alpha keys. These commands must be entered using the different *menus* on the calculator. The rest of the program can be entered by direct use of the keyboard. Note that each input line of the program must be followed by [**ENTER**].

Recall that we have just entered the program editor. To input the first line of the program, press [**PRGM**], highlight **I/O**, highlight **Disp** and press [**ENTER**]. Use the [**Alpha**] key to finish the first line of the program and press [**ENTER**].

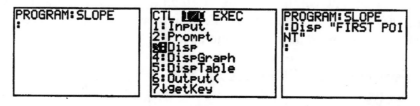

The second line of the program is entered in a similar manner. Press [**PRGM**], select **I/O** and press [**ENTER**]. Use the keyboard to complete the remainder of line 2 and press [**ENTER**].

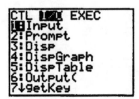 

The remainder of the program is input as above. Keep the following in mind:

- **Disp** and **Input** are found by pressing [**PRGM**], selecting **I/O**, high-lighting the required command, and pressing [**ENTER**].

- **If, Then, Else** and **End** are found by pressing [**PRGM**], selecting the required command and pressing [**ENTER**].

- Each command line (except the last one) must be followed by pressing [**ENTER**].

- The arrow keys can be combined with [**DEL**], [**INS**] and [**CLEAR**] to change lines in the program.

- Press [**QUIT**] when the input process is complete.

Once the **SLOPE** program has been entered into the calculator, it can be run by pressing [**PRGM**], scrolling to highlight the program name, and pressing [**ENTER**]. This will paste the program to the home screen. Press [**ENTER**] and the program will begin to execute.

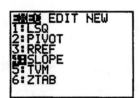 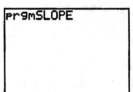 

To find the slope of the line connecting the points (–3.1, 5.2) and (4.3, –17.6), give the input below (follow each value by pressing [**ENTER**]) and let the program compute the slope.

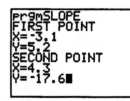 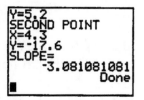

COIN and DICE: Two Simple Programs

This section contains two simple programs which might be useful for generating data and/or using with the Probability and Statistics chapters of your text. The **COIN** program simulates flipping fair coins. The user simply selects the number of coins to be flipped, and the program responds with the result. The **DICE** program simulates rolling fair *n*-sided dice. The user selects the number of dice and the number of sides on each die. The program responds with the results of a roll. Both programs make use of the built-in random number generator command in the TI-82/83/83+.

We start with the listing for the **COIN** program.

```
PROGRAM:COIN
:ClrHome
:ClrList L6
:Disp "PRESS ON TO QUIT"
:Disp "0=HEAD"
:Disp "1=TAIL"
:Input "NUM COINS=",N
:Lbl A
:For(I,1,N)
:iPart(2*rand)→L6(I)
:End
:Disp L6
:Disp "PRESS ENTER"
:Pause
:Goto A
```

The commands in the **COIN** program can be found in the following locations:

- **Input**, **Disp** and **ClrHome** are found by pressing [**PRGM**], high-lighting **I/O** and scrolling if necessary.

- **Lbl**, **For**, **Pause** and **Goto** are found by pressing [**PRGM**] (scroll, if necessary).

- **ClrList** is found by pressing [**STAT**].

- **iPart** is found by pressing [**MATH**] and highlighting **NUM**.

- **rand** is found by pressing [**MATH**] and highlighting **PRB**.

To execute the **COIN** program, press [**PRGM**], highlight **COIN** and press [**ENTER**] twice. Then follow the instructions on the screen. The output below shows the result of flipping 5 coins. The coins can be flipped again by pressing [**ENTER**]. Press [**ON**] to quit the program.

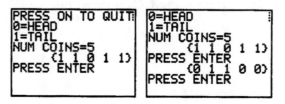

The **DICE** program is very similar. The listing is given below.

```
PROGRAM:DICE
:ClrHome
:ClrList L6
:Disp "PRESS ON TO QUIT"
:Input "NUM DICE=",N
:Input "NUM SIDES=",S
:Lbl A
:For(I,1,N)
:iPart (S*rand+1) →L6(I)
:End
```

:Disp L6
:Disp "PRESS ENTER"
:Pause
:Goto A

The bulleted items following the **COIN** program also apply to the **DICE** program.

To execute the **DICE** program, press [**PRGM**], highlight **DICE** and press [**ENTER**] twice. Then follow the instructions on the screen.

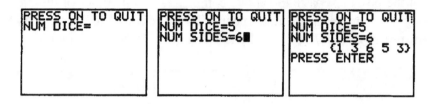

The **DICE** program will roll the dice again when [**ENTER**] is pressed. Press [**ON**] to quit the program.

LSQ: A Simple Least Squares Program

The program **LSQ** (given below) automates the process discussed in section 1.5. To use the program, simply enter the values for *x* and *y* into **L1** and **L2**, respectively, set the **GRAPH** window dimensions and execute the program.

```
PROGRAM:LSQ
:FnOff
:Func
:Plotsoff
:Plot1(Scatter,L1,L2,∘)
:LinReg(ax+b)
: "aX+b" → Y1
:DispGraph
```

To enter this program into either the TI-82 or TI-83/83+ calculator, press [**PRGM**] and use the arrow keys to select **NEW**. Press [**ENTER**] and input the name **LSQ** (note that the calculator is already in **ALPHA** mode) and press [**ENTER**].

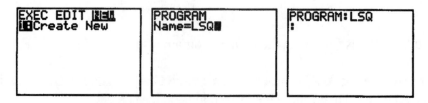

To input the program, paste the above commands into the lines of the program by accessing them in the appropriate menus. Then press [**VARS**], select **Y-VARS**, **On/Off**, press [**ENTER**], select **FnOff** and press [**ENTER**] again.

Now press [**MODE**], select **Func** and press [**ENTER**].

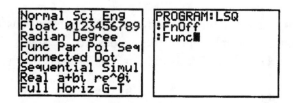

Press [**STAT PLOTS**], select **PlotsOff** and press [**ENTER**].

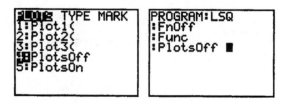

Again, press [**STAT PLOTS**] and press [**ENTER**].

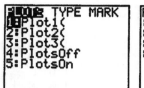 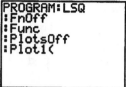

Return to [**STAT PLOTS**], select **TYPE** and press [**ENTER**]. Input **L1** and **L2** followed by commas.

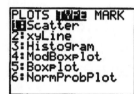 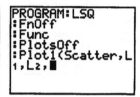

Return a final time to [**STAT PLOTS**], select **MARK** and press [**ENTER**] followed by a closed parenthesis.

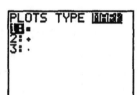 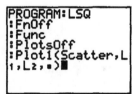

Press [**STAT**], highlight **CALC**, select **LinReg(ax+b)** and press [**ENTER**].

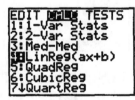 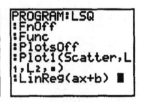

Insert double quotes, press [**VARS**], select Statistics, press [**ENTER**], highlight **EQ**, select **a** and press [**ENTER**]. Enter **X**, followed by a plus sign, followed by **b** (located in the same menu as a), insert a double quote and store the value in **Y1**.

 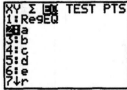 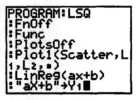

Finally, press [**PRGM**], highlight **I/O**, select **DispGraph** and press [**ENTER**].

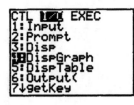 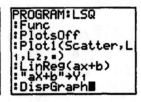

Press [**QUIT**] to exit the program editor and return to the home screen.

Let's use the **LSQ** program to solve the problem of finding the least squares line for the points

$$(1.1, -6), (2.3, -5.4), (5.4, -4.7), (7.3, -3.9), (8.1, -2.9)$$

Enter the data values into **L1** and **L2** and set the **GRAPH** window dimensions.

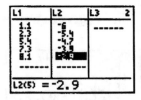 

Press [**PRGM**] and scroll to locate the **LSQ** program. Press [**ENTER**] to place the program on the home screen. Press [**ENTER**] again to execute the program.

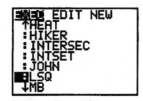  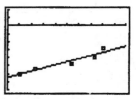

RREF: Automated Gauss-Jordan for the TI-82

The TI-83/83+ has a built-in program named **RREF** which performs the Gauss-Jordan method on an augmented matrix. The program **RREF** listed below will give the TI-82 the same capability.

Input the augmented matrix as [**D**] and run the **RREF** program. The result will be written to the screen and stored in the matrix [**E**].

```
PROGRAM:RREF
:dim([D] )→L6
:[D]→[E]
:L6(1)→M
:L6(2)→N
:1→S
:For(J,1,N)
:If S≤M
:Then
:0→K
:For(I,S,M)
:If abs [E](I,J)>10^(−8)
:Then
:rowSwap([E],I,S)→[E]
:I→K
:End
:End
:If K>0
:Then
```

:1/ ([E] (S,J))→R
:*row(R,[E],S)→[E]
:For(L,1,M)
:If L ≠ S
:Then
:*row+(– [E](L,J),[E],S,L)→[E]
:0→[E](L,J)
:End
:End
:1+S→S
:End
:End
:End
:[E]

Creating this program on the TI-82 is straight-forward, but the following comments might be helpful.

- **For, If, Then** and **End** are accessed by pressing [**PRGM**].

- **dim, rowSwap, *row** and ***row+** are accessed by pressing [**MATRX**] and highlighting **MATH**.

- **Disp** is accessed by pressing [**PRGM**] and highlighting **I/O**.

- [**D**] and [**E**] are accessed by pressing [**MATRX**].

- **L6** and **abs** are accessed from the keyboard.

Use this program to solve the system

$$\begin{pmatrix} x + 2y + 3z = 1 \\ 4x + 5y + 6z = 2 \\ 7x + 8y + 9z = 3 \end{pmatrix}$$

Enter the augmented matrix

$$\begin{pmatrix} 1 & 2 & 3 & 1 \\ 4 & 5 & 6 & 2 \\ 7 & 8 & 9 & 3 \end{pmatrix}$$

as matrix [**D**]. Run the program **RREF**. The result is written to the home screen and is stored in the matrix [**E**].

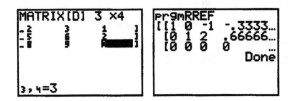

To view this matrix in fraction form, use the ▶**Frac** command.

The original system is equivalent to

$$\begin{pmatrix} x - z = -\dfrac{1}{3} \\ y + 2z = \dfrac{2}{3} \end{pmatrix}$$

where z is a free parameter.

Consequently, the system has infinitely many solutions given by

$$x = z - \frac{1}{3}, \; y = -2z + \frac{2}{3}, \text{ where } z \text{ is any real number.}$$

PIVOT: A Step-by-Step Simplex Method

The simplex method can be performed on either the TI-82 or TI-83/83+ by using the built in row operation commands. However, this process is very tedious. The program **PIVOT** (shown below) allows the user to simply identify the pivot element at each step. The rest of the work is done by the calculator.

This program can be used on either the TI-82 or TI-83/83+. The only difference occurs on line 21 with the **abs** command. The line shown in the program listing below corresponds to the appearance on the TI-83/83+. The corresponding line on the TI-82 is given by

$$\text{:If abs [D](K,L)} < 10^{\wedge}(-10)$$

To use this program, input the initial simplex table as matrix **[D]** and run the program. Note that the updated simplex table is stored after each step in the matrix **[E]**, and the original simplex table **[D]** is unchanged. Consequently, if you make a mistake choosing a pivot element, you can simply restart the program.

```
PROGRAM:PIVOT
:[D]→[E]
:Lbl A
:Menu("MENU","PIVOT",B,"QUIT", C)
:Lbl B
:ClrHome
:Disp "ENTER TO CONT"
:Pause [E]▶Frac
:Input "ROW=",R
:Input "COL",C
:[E] (R,C)→M
:*row(M⁻¹, [E],R)→[E]
:dim([E] )→L6
:For(K,1,L6(1))
:If (K≠R)
:Then
```

108

```
:-[E](K,C)→M
:*row+(M,[E],R,K)→[E]
:End
:End
:For(K,1,L6(1))
:For(L,1,L6(2))
:If abs([E](K,L))<10^(-10)
:Then
:0→[E](K,L)
:End
:End
:End
:Disp "ENTER TO CONT"
:Pause [E]▶Frac
:Goto A
:Lbl C
```

Almost all of the commands used in this program were also used in the **RREF** program. Refer to the bulleted items in the previous section for their locations on the calculator. Exceptions are shown below:

- **Lbl**, **Menu**, **Pause** and **Goto** can be obtained by pressing [**PRGM**] and scrolling if necessary.

- **ClrHome** and **Input** can be accessed by pressing [**PRGM**], highlighting **I/O** and scrolling if necessary.

See section 4.2 for an example using the **PIVOT** program.

TVM: Time Value of Money for the TI-82

The **TVM** program gives the TI-82 the capabilities that can be found in the **TVM Solver** on the TI-83/83+. The program is listed below.

```
PROGRAM:TVM
:"F-( R((1+I)^(N+B)-1)/I)+P(1+I)^N"→Y1
```

```
:Lbl Z
:ClrHome
:Input "N=",N
:Input "I=",J
:Input "PV=",P
:Input "PMT=",R
:Input "FV=",F
:Input "C/Y=",Y
:Menu(" PAYMENT"," END" E," BEGIN",B)
:Lbl E
:0→B
:Goto A
:Lbl B
:1→B
:Lbl A
:J/(100*Y)→I
:Menu("SOLVE FOR" ,"N",N,"I",J,"PV",P,"PMT",R,"FV",F)
:Lbl N
:solve(Y1,N,100,{0,100000})→N
:Disp N
:Goto H
:Lbl J
:solve(Yl,I,.1,{0,3})→I
:Y*I*100→J
:Disp J
:Goto H
:Lbl P
:solve(Y1,P,0,{−(10^50),10^50})→P
:Disp P
:Goto H
:Lbl R
:solve(Y1,R,0,{−(10^50),10^50})→R
:Disp −R
:Goto H
:Lbl F
:solve(Yl,F,0,{−(10^50),10^50})→F
:Disp F
```

```
:Lbl H
:Disp "ENTER TO CONT"
:Pause
:Menu("CHOOSE AN OPTION"," TVM",Z,"QUIT",Q)
:Lbl Q
```

The **TVM** program includes all of the features of the TI-83/83+ **TVM Solver** except **P/Y**, which is assumed to be the same as **C/Y**. The specific values that **TVM** can compute are given below.

- **N** is the number of payment periods

- **I %** is the annual interest rate given as a percentage (i.e. 4% is input as 4)

- **PV** is the present value

- **PMT** is the payment amount

- **FV** is the future value

- **C/Y** is the number of conversion periods per year

- **PMT: END BEGIN** is when the payments occur in a payment period

To compute one of these values, enter each of the other values, but place a "dummy" value for the value to be computed. For example, to compute **PMT** corresponding to **N = 200, I = 4, PV = 0, FV = 80,000, C/Y = 12** and payments at the end of each period, we execute the **TVM** program and make the choices shown below. Note that the value given for **PMT** is a "dummy" value. Press **[ENTER]** after selecting **END**.

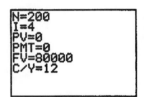

Now select **PMT** and press **[ENTER]**.

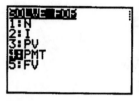 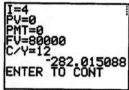

The payment amount is $282.02 per month.

ZTAB: Z Scores and Areas for the TI-82

This program can be used to find the area under a normal curve between two x values, and it can be used to find the x value given the area under the normal curve to the left of the x value.

PROGRAM:ZTAB
:"1/$\sqrt{(2\pi)}$*e^(–X^2/2)"$\rightarrow$Y1
:ClrHome
:Input "MEAN=",M
:Input "STD DEV=",S
:Lbl A
:Menu("ZTAB","AREA",B,"Z AND X VALS",F,"QUIT",E)
:Lbl B
:Menu("TAILS?","1 TAIL",C,"2 TAIL",D,"QUIT",E)
:Lbl C
:Input "X=",X
:(X–M)/S$\rightarrow$Z
:Disp "CORRESPONDING Z"
:Disp Z
:Disp "AREA"
:1/2+fnInt(Y1,X,0,Z)$\rightarrow$A
:Disp A
:Disp "PRESS ENTER"
:Pause
:Goto A
:Lbl D
:Input "LOWER X=",L

```
:Input "UPPER X=",U
:(L–M)/S→Y
:(U–M)/S→Z
:Disp "CORRESPONDING Zs"
:Disp Y,Z
:Disp "AREA"
:fnInt (Y1,X,Y,Z)→A
:Disp A
:Disp "PRESS ENTER"
:Pause
:Goto A
:Lbl F
:Disp "AREA TO THE"
:Disp "LEFT OF THE PT"
:Input A
:Lbl G
:0→Z
:1/2+fhInt(Y1,X,0,Z)–A→E
:While abs(E/A)>0.0001
:Z–E/Yl(Z)→Z
:1/2+fnInt(Yl,X,0,Z)–A→E
:End
:M+Z*S→X
:Disp "X=",X
:Disp "Z=",Z
:Disp "PRESS ENTER"
:Pause
:Goto A
:Lbl E
```

Example: Suppose X is a normal variable with $\mu = 3$ and $\sigma = 5$. We can use the **ZTAB** program to compute the following:

a. $P(2 \le X \le 7)$

b. $P(X \le 8)$

c. x so that $P(X \le x) = 0.8$

Solution: To execute **ZTAB**, press **[PRGM]**, highlight **ZTAB** and press **[ENTER]** twice. Input the mean and the standard deviation into the calculator as shown below.

For part a, note that computing $P(2 \le X \le 7)$ is equivalent to finding the area under the normal curve between $x = 2$ and $x = 7$. Consequently, we select **2 TAIL**, press **[ENTER]** and supply **LOWER X** and **UPPER X**. The program computes the corresponding z values and the area.

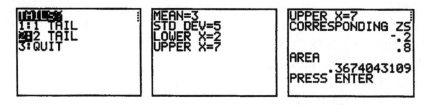

As a result, $P(2 \le X \le 7)$ is approximately 0.3674.

For part b, press **[ENTER]** to return to the **ZTAB** screen, select **AREA** and press **[ENTER]**. This gives the first screen below. Select **1 TAIL**, press **[ENTER]** and input the value for x shown. The calculator responds with the corresponding z value and the area to the left of x.

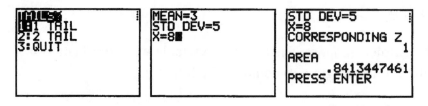

Therefore, $P(X \le 8)$ is approximately 0.8413.

Finally, for part c, return to the **ZTAB** screen as above. Select **Z AND X VALS**, press [**ENTER**], and input the area to the left of the unknown *x*. The calculator returns the *x* value and the corresponding *z* value.

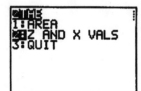 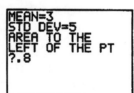